The History of the International Biometric Society

Lynne Billard

The History of the International Biometric Society

CRC Press is an imprint of the
Taylor & Francis Group, an **informa** business
A CHAPMAN & HALL BOOK

First edition published 2022
by CRC Press
6000 Broken Sound Parkway NW, Suite 300, Boca Raton, FL 33487-2742

and by CRC Press
4 Park Square, Milton Park, Abingdon, Oxon, OX14 4RN

CRC Press is an imprint of Taylor & Francis Group, LLC

ISBN: 9781032258645 (hbk)
ISBN: 9781032258669 (pbk)
ISBN: 9781003285366 (ebk)

DOI: 10.1201/9781003285366

Typeset in Palatino
by codeMantra

Contents

Foreword

The International Biometric Society (IBS) is this year celebrating the seventy-fifth anniversary of its establishment. In his Presidential Address at the 2006 International Biometric Conference in Montreal, Professor Tom Louis said that having an appreciation of our history as it is being created (our future as history) helps make better sense of the present and to plan more wisely for the future. Indeed, time goes by quickly and, before we know it, our present has become our history and our future has become the present!

Since the Society's birth in 1947 and the time leading up to it, copious archival materials have accumulated, representing paper records, letters, notes and old journals. When the Society made some changes to its management structure in 1994, these materials found a temporary home in the office of Professor Lynne Billard at the University of Georgia. Lynne had served as IBS President from 1994–1995 and had stayed very active in the Society subsequently. The materials sat quietly, undisturbed in several file boxes in the corner of Lynne's office for a number of years until a conversation in 2010 with then President Kaye Basford, who suggested that it would be great if someone could go through the archives and write up an account of IBS' history. Some material has of course been published during the years in various places, including several articles in the journal *Biometrics*. Further details can be found in the "Our History" tab of our Society's webpage. But a full account had never appeared, certainly not with all the rich detail about the formation of the various regions and so on. Lynne generously agreed to take on the task and during the course of the next few years, painstakingly worked her way through the materials, making notes, gathering photographs and putting it all together in a fascinating story. The effort has finally come to fruition in the form of this monograph!

The IBS is truly grateful to Lynne for all the hard work she has invested in creating this book! I hope you will agree with me that it provides interesting, indeed invaluable insights into the personalities and the global movements that have shaped our Society during the years. As if donating her time was not enough, Lynne has arranged that any royalties generated from the publication will go back to the IBS to support its travel grant program.

I hope you will enjoy reading Lynne's book as much as I did!

Sincerely
Professor Geert Verbeke
President of the International Biometric Society

Foreword

Preface

Boxes of the archives of the International Biometric Society (Society) had found a home gathering dust in my office. Eventually, it came time to view the contents. By happenstance, the first folder pulled from these boxes consisted of the Minutes of the meeting at which the Society was founded on Saturday morning September 6 1947 at Woods Hole. This was, or so it seemed to me, pure gold-dust. Later when rifling through other boxes, a folder on Fisher's death popped up; inside were letters written to Bliss and Fisher's son with details (another startling find). By now, it was abundantly clear that these boxes were a treasure trove about the Society. It was also apparent that, rather than being locked away, these materials had to be prepared into some format for sharing. Thus it came to be that I started to read, to digest, to summarize all these pages buried in this gold-mine. There have been many enjoyable and unexpected moments as this road was travelled.

One highlight was the sense of gaining great insights into the personalities of our early leaders. Communications were by letters; phones were expensive, email and related computer technologies did not exist; so letters served as the primary source of information. Some say history can be as "dry bones." Not these pages - the exchanges sparkled with wit, grace, concern, pragmatism, warmth and care - especially care for the growth and impact of the Society itself. At times, I felt I was sitting in a member's cosy living room listening, even being entertained by reminiscences of experiences (and trials, yes) as the Society was nurtured. What a privilege it was! My hope is that you can feel their pride in their work as you read my summary of events.

Another pleasant surprise concerned the photos. This history contains lots of photos. Letters seeking permission to use said photos typically brought immediate positive responses; that was gratifying. But what was unexpected was that on their own initiative most responders sent higher quality photos (I had included my version in the request letter, along with reasons why their person was important to the story). In their unique ways, these folk were conveying their own excitement in the preparation of this work. Of course, even the "best" photos of days past do not match today's quality – it is after all, a History! Earlier, finding these photos was another adventure, a journey which inevitably led me to numerous articles about our distinguished forebears. To tell of just one - an early Italian Region regional president Gustavo Barbensi once published a volume of Shakespeare sonnets translated into Italian. Although not in person, I certainly met Barbensi along my journey.

May you too enjoy meeting our early members as you read!

Lynne Billard

Author

Lynne Billard received a First Class Honors BS degree in 1966 and a PhD degree in 1969 both from the University of New South Wales Australia. She is a former department head of statistics and computer science; since 1992, she has been University Professor in the Department of Statistics at the University of Georgia USA. Billard is a former international President of the International Biometric Society (IBS, 1994-1995), regional president of ENAR (1985), and President of the American Statistical Association (ASA, 1996). She is the recipient of the Wilks Award (ASA career research), Carver Award (Institute of Mathematical Statistics, IMS service), Founders Award (ASA service), Elizabeth Scott and FN David COPSS Awards, Janet Norwood Award, University of New South Wales Alumni Award, among many other distinguished honors. She has served on innumerable scientific boards including the National Academy of Sciences' Board of Mathematical Sciences, US Secretary of Commerce Census 2000 Advisory Committee, and as Chair of the Conference Board of Mathematical Sciences, as well as associations such as IBS Council, International Statistical Institute (ISI) Council, and IMS Council. She has over two hundred and fifty publications and ten books. Her research has drawn from the fields of sequential analysis, epidemic theory (including AIDS-HIV), time series, and most recently symbolic data analysis. She is a Fellow of the ASA, IMS, and the American Association for the Advancement of the Sciences, and an elected Member of the ISI.

Gallery of Presidents

1947-1949
Sir Ronald Aylmer Fisher

1950-1951
Arthur Linder

1952-1953
Georges Darmois

1954-1955
William G Cochran

1956-1957
E Alfred Cornish

1958-1959
Cyril H Goulden

1960-1961
Léopold Martin

1962-1963
Chester I Bliss

1964-1965
David J Finney

1966-1967
Luigi L Cavalli-Sforza

1968-1969
Gertrude M Cox

1970-1971
Berthold Schneider

1972-1973
Peter Armitage

1974-1975
C R Rao

1976-1977
Henri L Le Roy

1978-1979
John A Nelder

1980-1981
Richard M Cormack

1982-1983 Herbert A David
1984-1985 Pierre Dagnelie
1986-1987 Geoffrey H Freeman
1988-1989 Jonas H Ellenberg
1990-1991 Richard Tomassone
1992-1993 Niels Keiding
1994-1995 Lynne Billard
1996-1997 Byron J T Morgan
1998-1999 Susan R Wilson
2000-2001 Nanny Wermuth
2002-2003 Norman E Breslow
2004-2005 Geert Molenberghs
2006-2007 Thomas A Louis
2008-2009 Andrew Mead
2010-2011 Kaye E Basford
2012-2013 Clarice G Borges Demetrio

2014-2015 2016-2017 2018-2019 2020-2021 2022-2023
John P Hinde, Elizabeth A Thompson, Louise M Ryan, Geert Verbeke, José C Pinheiro

1

Introduction

By all accounts, Chester Ittner Bliss[1] was the spark that led to the formation of the Biometric Society, later called the International Biometric Society (the "Society"). Ronald Aylmer Fisher (from 1952, Sir Ronald) and Gertrude Mary Cox had key supporting roles (Figure 1.1), along with numerous others of our past statistical luminaries. In recognition of the fiftieth anniversary of Fisher's death, Billard (2014) focused on his role in the origins of the Society. By drawing upon the archival records, we look at the history of the Society across all Society dignitaries, without any undue emphasis on any one person. However, certainly, the vitally important role played by Bliss could not be ignored, but as the history unfolded, clearly many more of our predecessors also contributed in important ways.

The archives consist of letters, and then more letters, covering proposals, thoughts, and actions from the various actors. Outside of conferences, letters were the primary medium for communication, though even then there were extensive records of deliberations. It is extraordinary how deeply committed the members were to the Society, at least as revealed by the archives. Even when voting (on whatever issues), there would frequently be added comments, some pages long; Bliss always faithfully recorded all opinions in his subsequent reports. In this volume, we focus on issues and events that engaged the attention of the officers at the Society, i.e., international, level. Some of these records are riveting, entertaining, intriguing, and colorful, while some issues were difficult to handle (but even these often resulted in constitutional or procedural changes that benefited the Society). For most, it was the President and/or Secretary carrying the banner seeking resolution. Bliss, David Finney, and Cox were "brilliant" (to use hyperbole), despite Finney's self-deprecating remark to an incoming Secretary that with time you will discover "I am a thorn in the flesh" and his handwritten promise (when informed of his election as Vice-President) that "I shall endeavour to be useful." In reality, however, a distinguishing feature throughout was the thoughtfulness, diplomacy, tact, and an ability to rise "above the fray" to look at issues from the broader perspective. Most officers made lasting and important contributions to the betterment of the Society. Therefore, in this history, we attempt to convey the details and flavor of events that unfolded

[1]Full names are given at the first mention of a person within a chapter; subsequently, only the last name is used.

DOI: 10.1201/9781003285366-1

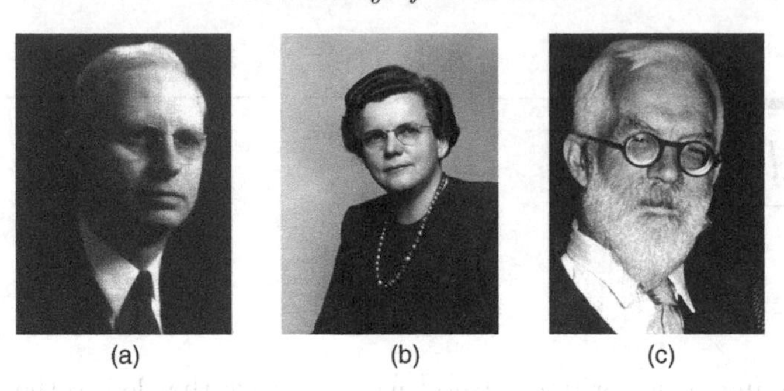

FIGURE 1.1
(a) Chester I. Bliss, (b) Gertrude M. Cox, and (c) Ronald A. Fisher.

over the years as depicted in the archives rather than the sanitized versions of published Society reports.

In addition, the archives contain numerous letters detailing possible nominations for a wide variety of positions along with sometimes quite frank assessments of suitability, address changes of members (providing an interesting tracking of institutional moves of members for anyone so inclined to follow these), name changes due to marriage, notices that a journal issue had not arrived, letters to members who were in arrears with their dues payments (who knew?!) as well as regional delinquency rates in collecting dues, requests for membership forms, budgets, records of all kinds, and so forth. There are also stacks of records of voting returns complete with commentary sometimes about the issues involved. Except where pertinent to the story, this present history has not reported on any individual letter, notice, or voting record. What shines through however, is the pride of being a member, a builder, and a conservator, of what has become a very important Society especially for the biometrical world to which it was addressed.

Therefore, we trace the formation of the Society (in Chapter 2) as reported and described in the voluminous archive records. This takes us back to the beginnings which culminated in the Woods Hole International Biometric Conference meeting (September 5–6, 1947) at which the Society was formally established. Then, in Chapter 3, we follow significant markers that defined the Society down through subsequent years; these include memberships, committees, or simply ‘events’ that captured the attention of the leadership of the day. Committees tended to be added with the emergence of new undertakings and certain needs which required attention; these were established permanently as standing committees or as temporary ad hoc committees. That chapter also includes details of international affiliations.

From the outset, the Society was to be an international society rather than a federation of national associations. Thus, there was a unique regional structure spanning global geographical areas, each with its own autonomy but each very much a part of the whole Society. The structure and formation of these regions, and also national groups for smaller clusters of members, is outlined in Chapter 4. That chapter includes details relating to the first four regions, specifically, the Eastern North American Region (ENAR), the British Region (BR), the Western North American Region (WNAR), and the Australasian Region (AR), all formed within one year of the establishment of the Society. Then, in Chapter 5, aspects of those regions that formed over the next ten years are described. These include the Indian Region (IR) in 1949 (though this region collapsed in 1951, moved to group status in 1953 before returning to regional status again in 1989), the Région Française (French Region, RF) in 1949, the Société Adolphe Quetelet was formed as the Belgian Region (RBe) in 1952, the Italian Region (RItl) in 1953, the Deutsche Region (German Region, DR) in 1955, and the Região Brasileira (Brazilian Region, RBras) in 1956. Those regions that formed over subsequent years are covered in Chapter 6. Here, we learn that Afdeling Netherland (The Netherlands Region, ANed) although a national group as early as 1949 became a region in 1960, Region Österreich-Schweiz (Austro-Swiss Region, ROeS) which evolved from an earlier Swiss National Group came into existence in 1962, the Japanese Region (JR) formed in 1979, the Nordic Region's (NR) emergence in 1982 consisted of national groups making up the Nordic geographical region and merged with the Baltic National Group in 2002 to become the Nordic-Baltic Region (NBR), the Hungarian Region was established in 1988 though it later reverted to national group status (in 2003), and the Spanish Region (REsp) was approved in 1992. Chapter 6 also describes developments in Eastern Europe; as well as national groups not elsewhere covered and the formation of networks (usually to facilitate scientific meetings of smaller national groups). Abbreviations of Regions and National Groups are provided in Chapter 12, Table 12.1.

The core of the Society, indeed its scientific backbone, is the journal *Biometrics*; later other publications were added. Thus, the role of Society publications is outlined in Chapters 7 and 8. Scientific meetings, be these International Biometric Conferences (IBC) or Symposia, are also an important part of the scientific mission. These are described in Chapter 9.

The Society is governed by the Executive Committee consisting of the international Society President ("President"), the international Society Secretary ("Secretary"), international Society Treasurer ("Treasurer"), and the *Biometrics* Editor (from 1956; later, with the addition of new journals, this role was filled by an editorial representative variously defined over the years). This Executive Committee worked with the International Society's Council ("Council"). An over-riding criterion of the structure of Council was that all regions be represented; Council members were the conduit for information to pass to and from the international Society level and the regional and national group levels. The Society functioned through its Constitution and By-laws. Each region also had its own governing processes and its own By-laws (which minimally had to comply with those of the Society). As the Society expanded in membership and regions, as various issues arose, and as new roles and endeavors were taken on, it would become necessary to revise these entities. The fundamentals of the Constitution and its changes are described in Chapter 10. That chapter also describes the evolution of the business and management aspects of the Society.

Chapters 1–10 constitute detailed histories in the Society's first fifty years, up to 1997. However, in Chapter 11, some highlight markers that have occurred since 1997 are briefly outlined. Occasionally, when adjudged necessary for purposes of continuity of the narrative, a snippet of post-1997 detail is included in an earlier chapter where applicable.

Finally, for the record, a listing of the former Presidents, Secretaries, and Treasurers is provided in Chapter 12. This chapter also lists former regional presidents and national group secretaries, and editors, as well as table 12.2 summarizing locations and local and program chairs for the International Biometric Conferences and key Symposia.

The official face of any association, including our Society, is its leader, its international President; therefore, this volume starts off with a Gallery of Presidents displayed in the beginning. However, oftentimes it is the international Secretary well supported by the international Treasurer both of whom are the bedrock on which Society activities and communications are built; thus, this volume closes with a Gallery of Secretaries and Treasurers.

2

In the Beginning – Woods Hole

The International Biometric Society ("Society") was founded at Woods Hole on September 6, 1947. This chapter discusses the formation of the Society through the eyes of the Archival records, focussing on details of events leading up to and at Woods Hole.

In the American Statistical Association's (ASA) new constitution of 1938, provision was made for the establishment of sections. Thus, immediately its first section, the Biometrics Section ("Section"), was formed on March 14, 1938, for members interested in aspects biometry applied to the biological sciences. Then, in 1945, because no meetings could be held due to World War II, the Section started *Biometrics Bulletin* with Gertrude Cox as Editor, to serve as a newsletter to keep Section members informed of events.

The ASA constitution also allowed for the "growth of sections into self-supporting associate societies" as part of ASA's program. Therefore, in 1945, Chester Bliss as Section Chair, appointed a committee to draft up a constitution which if adopted would become "Biometric Society, an Associate of the [ASA]", i.e., an American Biometric Society somewhat akin to what the economists had done. In the end, at its January 1946 meeting, these plans were tabled. By now, Cox was having a difficult time with the ASA office over publications issues (even saying "I am floored with this editorial set-up"), culminating with her long October 3, 1946 letter to her editorial board, conveying her concerns. Likewise, Bliss also in frustration at being "a ward of the A.S.A. Board of Directors with no authority in its own right" wrote a five-page letter on October 7, 1946 to Section officers, outlining some proposals for the journal (such as making it a quarterly journal "at least as good as the *Journal of the American Statistical Association*") and importantly proposals for what was a "call" for a new society. " ...[W]e can [not] go through another year on the present basis" concluded Bliss. Both letters generated a flurry of responses, in support.

Meantime, the International Statistical Institute (ISI), an international organization whose prime focus had been government statistical entities since its 1885 formation, under its new constitution had expanded its scope to include non-government members including mathematical and biometrical statisticians. The first ISI session after this expansion was to be held in Washington, DC in September 1947. When the preliminary scientific program was published in early 1947, Bliss among many others felt that the biostatistical community had been ignored. Fortuitously, Bliss encountered

DOI: 10.1201/9781003285366-2

the economist Charles Frederick Roos on a Saturday March 29 train trip from New York to Princeton; on bemoaning the ISI development, Roos replied that maybe Bliss should follow the route taken by the economists and form their own international organization. One aspect of tabling the earlier proposal for an American Biometric Society had been a sense that too much effort would be wasted on a purely Section organization; suddenly, Bliss felt he had found "the missing piece" (Bliss, 1958), that an "international biometric society ... offered prospects not available to [the Section ...]." He was galvanized into action, even discussing the idea with John Wilder Tukey among others while still at Princeton.

Thus, back at his home office, on Monday March 31, a scant two days later, Bliss wrote to the Section Chair Dan DeLury with his proposals. DeLury replied the following week April 8, with a framework for proceeding and an Organizing Committee (Bliss as Chair, Edwin deBeer, Horace Norton, and Tukey) to effect the implementation. Taking his cues from Bliss' letter, DeLury specifically asked the Committee to "(1) Issue a call for the organization of an international biometric society. (2) Make arrangements for an organizational meeting (time, place, etc.) and issue invitations to foreign and American biometricians to attend. (3) Prepare a report to the Board of Directors of the A.S.A. informing them of this undertaking. [and] (4) Prepare a provisional draft of a constitution to serve as a starting-point for the organizational meeting." A handwritten addendum to DeLury's letter referred to Bliss' "your" idea, asked Bliss if his (DeLury's) call here was what Bliss wanted, and opined that "An international organization is needed and may well be built upon the start that the B. S. [Biometrics Section] has made." It is very clear that Bliss was the key person. Even so, ever the true statesman that he was, ever mindful of the mighty significance of Fisher as a statistician/scientist, Bliss engaged and consulted with his great friend Ronald Fisher throughout, asking/letting Fisher play significant roles whenever appropriate, judging by Fisher's April 14, 1947 letter to Bliss with his reference to "your committee." Later, on May 30, 1947, Fisher tells Bliss that "I think your plans are very good ... ." However, it was Bliss himself who orchestrated events in establishing the Society.

Letters flew hither and yon as ideas were sought and digested. A wide range of opinions was sought; while there were a couple of cautions against starting new organizations so soon, most respondents were enthusiastic. Eventually, there was convergence toward having a meeting at Woods Hole as this was an important biological research center; the meeting would actually be a conference with some scientific sessions (with foreign speakers who then should be in a better position to obtain funds from their home countries to attend). The Organizing Committee met on May 10–11 to brainstorm and decided to proceed. In a set of June 10 letters, the Committee invited Albert Francis Blakeslee, William Gemmell Cochran, Haldan Keffer Hartline, Nicolas Rashevsky, and John von Neumann as members; Edmund W Sinnott and Edwin B Wilson were added as National Research Council (NRC) representatives by the NRC Chair Detlev Wulf Bronk. Tukey took the lead on drafting a constitution. He had a copy of the January 1946 Section draft; thus, letters to various folk sought information as to whether a more recent draft existed and if so where. He submitted a potential constitution to Bliss on May 15, 1947, one modeled after that of the Econometric Society. Later, on June 20, 1947, Bliss consulted with the Organizing Committee about calling the upcoming conference the "First International Biometric Conference" (agreed).

Action was now at a frenetic pace judging by the archival records. Lists of people to invite to the Conference were being drawn up. The ISI was particularly helpful, making available its list of attendees for its Washington, DC session, further inviting additional people to its own meeting, as suggested by Bliss. It was thought that if folk had expenses to the ISI session, then the cost of adding the extra step to attend Woods Hole would be easier. For American delegates, the Institute of Mathematical Statistics (IMS) had a meeting scheduled for New Haven for August 30 to September 3. By July 3, 207 invitations to potential participants in 20 countries had been issued, by airmail; letters to American invitees were sent by regular surface mail on August 4. In addition, in a June 12 discussion with Warren Weaver (one of its directors), Bliss was advised that $1000 from the Rockefeller Foundation would be forthcoming (but channeled through the Marine Biological Laboratory at Woods Hole) to cover train fares of foreign delegates from New York to Woods Hole; this grant was also available to cover local organizational expenses.

To stress that the proposed organization was to be international and not an American entity, "foreigners" were invited to give scientific talks. Thus, arrangements were made so that on the Friday morning, after the preliminaries that set the stage for the unfolding events, Fisher would give a presentation on quantitative genetics, and the afternoon session would deal with reports on the scientific developments of biometry in various countries. The Saturday afternoon featured a talk on quantitative growth, in French, by Georges Teissier (France). The Conference itself ended with closing remarks by Teissier in French. See Figure 2.1.

The Organizing Committee, expanded to include several foreign members, met Wednesday evening September 3, 1947 at Bliss' Yale office in New Haven, to fine-tune plans (including Tukey's draft constitution all ready to be discussed and adopted at Woods Hole itself). When they departed for Woods Hole, Bliss famously said "our 'homework' had been done."

The Saturday morning September 6 was devoted to this discussion. First, however, the assembly had to set in motion the events leading to that momentous occasion. Therefore, after an opening greeting on the first morning Friday September 5, Bliss reported that he had asked a committee (Arthur Linder, Carl Fredrick Kossack, and Ernest Walter Lindstrom) to nominate a Conference Committee; Chair Teissier, Co-Chair Fisher, and Secretary Bliss were so nominated and duly elected. Then, Bliss gave his report from the Organizing Committee; this report outlined the background and events that led up to that day's activities. Bliss then moved that the

PROGRAM

Friday, September 5

9:30 A.M. OPENING SESSION
Welcome to the Marine Biological Laboratory, CHARLES L. PACKARD, Director
Election of permanent officers of the Conference

10:00 A.M. QUANTITATIVE GENETICS
Chairman: A. F. BLAKESLEE, Smith College
"A Quantitative Theory of Genetic Recombination", R. A. FISHER, Cambridge University
Discussion opened by D. G. CATCHESIDE, Cambridge University

2:00 P.M. RECENT BIOMETRIC DEVELOPMENTS OVERSEAS
Chairman: E. B. WILSON, Harvard University
Informal reports by G. RASCH, Copenhagen;
O. TEDIN, Svalof;
R. C. BOSE, Calcutta
and others.

Saturday, September 6

10:00 A.M. INTERNATIONAL COOPERATION IN BIOMETRICS
Chairman: C. E. DIEULEFAIT, Rosario, Argentina
General Discussion

2:00 P.M. QUANTITATIVE GROWTH
Chairman: LESLIE F. NIMS, Brookhaven Laboratories
"La Relation d'Allométrie, sa Signification Statistique et sa Logique"
G. TEISSIER, Centre National de la Recherche Scientifique, Paris.
Discussion opened by JACQUES MONOD, Institute Pasteur, Paris.

FIGURE 2.1
Scientific program – IBC 1, September 5–6, 1947, Woods Hole.

Chair name a Committee on International Organization to consider what type of organization would be suitable to advance biometry and to prepare a draft resolution. Thus, at the beginning of the afternoon session, the assembly were told that Maurice Henry Belz (Australia), Bliss (USA), Raj Chandra Bose (India), Bronk (USA), Carlos Eugenio Dieulefait (Argentina), Fisher (Chair, UK), John William Hopkins (Canada), Linder (Switzerland), M G Neurdenburg (The Netherlands), Georg William Rasch (Denmark), Teissier (France), and Tukey (USA) had been appointed. This Committee met that evening (presumably after the clam bake, given the Minutes reported that it started at "about 10:00 P.M." and adjourned "after 1:00 A.M."); their recommendations were prepared, and their revised draft constitution was mimeographed, to be presented the next morning.

The Conference Chair Teissier opened the Saturday September 6 morning session "International Cooperation in Biometrics" at 10:20 A.M., and after announcements, he turned the session over to Dieulefait. Dieulefait (who was active in the ISI) spoke of the reorganization of the ISI and suggested that "the society that might be formed this morning" consider affiliation with ISI. Then Fisher submitted the report of the Committee on International Organization. This Committee recommended the formation of "an international membership organization" whose draft constitution was then circulated.

Belz took over as Chair; and so began the debate of the proposed constitution "article by article." Clearly, before discussing the constitutional components, an initial debate over the name of this organization ("Title" in the Minutes) had to be undertaken. A major point was whether or not the word "international" should be part of the name; there were arguments for and against. Other arguments concerned whether or not the word "biometric" should end with an "s". Brevity won the day. In fact, Tukey had argued for both these points in an April 14, 1947 letter to Bliss. In the end, (from the Minutes) "*Belz.* The name of the organization shall be the 'Biometric Society' and a short description clarifying the objective of the Society shall appear on the letterheads. The motion was seconded .... [including Belz' subtitle]." That done, the assembly could now move on to discussing the constitution. A portion of the morning discussion is shown in the Minutes of Figure 2.2.

The detail of the debates is pages long, sometimes fraught with "considerable feeling one way or the other," but always productive. It was also sometimes hilarious, as is seen in Figure 2.3. For example, the Minutes, when debating the first item "1. Scope of the Society," recorded: "*Geiringer*. Omit the word 'effective' in the third line. *Tukey*. leave word 'effective' in. *Rashevsky*. remove word 'effective'. *Hotelling*. pointed out that word 'effective' had an effective meaning". The word stayed in. Discussions of a particular word or phrase, its use, its place in a sentence, and so forth, were all extensively debated. This close and determined attention to the entire draft went on all morning.

The tenures of President and Vice-President[1] were quickly set at a maximum of two consecutive one-year terms. The term of the Secretary and Treasurer officers positions received considerable attention. On the one hand, it could be advantageous to have some permanent officers to ensure smoothness in routine operations; on the other hand, e.g., it could be difficult to remove a "bad secretary ... tactfully," that it was important that someone not "lose face," and that Council "might feel shy in firing an undesirable officer." A term limit of six years (e.g.), or two consecutive three-year terms, would make it difficult for removal, particularly since it was felt that the

[1] Until 1953, a Vice-President was effectively a regional president; this is distinct from the Society Vice-President position that started in 1963.

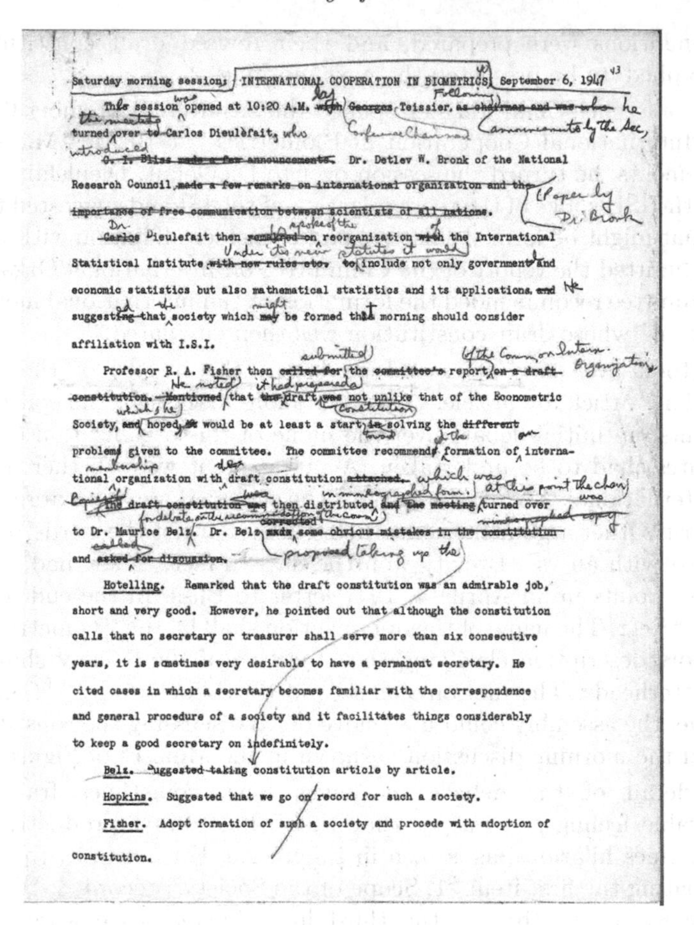

Saturday morning session; INTERNATIONAL COOPERATION IN BIOMETRICS September 6, 1947

This session opened at 10:20 A.M. with Georges Teissier, as chairman and was who turned over to Carlos Dieulèfait, who

C. I. Bliss made a few announcements. Dr. Detlev W. Bronk of the National Research Council made a few remarks on international organization and the importance of free communication between scientists of all nations.

Carlos Dieulefait then remarked on reorganization with the International Statistical Institute with new rules etc. to include not only government and economic statistics but also mathematical statistics and its applications, and suggesting that society which may be formed this morning should consider affiliation with I.S.I.

Professor R. A. Fisher then called for the committee's report on a draft constitution. Mentioned that the draft was not unlike that of the Econometric Society, and hoped it would be at least a start in solving the different problems given to the committee. The committee recommends formation of international organization with draft constitution attached.

The draft constitution was then distributed and the meeting turned over to Dr. Maurice Belz. Dr. Belz made some obvious mistakes in the constitution and asked for discussion.

Hotelling. Remarked that the draft constitution was an admirable job, short and very good. However, he pointed out that although the constitution calls that no secretary or treasurer shall serve more than six consecutive years, it is sometimes very desirable to have a permanent secretary. He cited cases in which a secretary becomes familiar with the correspondence and general procedure of a society and it facilitates things considerably to keep a good secretary on indefinitely.

Belz. Suggested taking constitution article by article.

Hopkins. Suggested that we go on record for such a society.

Fisher. Adopt formation of such a society and procede with adoption of constitution.

FIGURE 2.2
Minutes Saturday morning session, with Bliss' pencilled edits.

default for an automatic extension to a second term was the more likely action. Thus, it was finally agreed that each term was one year with no limit on the number of terms. As an aside, Neurdenburg's suggestion that there should be "a provision for keeping and for getting rid of a Secretary" proved providential in light to his own actions as National Secretary of The Netherlands (see Chapter 6, Section 6.1).

The last record came from "*Kossack*. Amendment to Tukey's motion [about Charter members]. [That] the drafting Committee [i.e, Fisher's Committee on International Cooperation] be the first Council and have power to delegate other members to Council." "Motion seconded and carried as amended. Constitution adopted as amended." The final version of the First Constitution is shown in Figure 2.4. Elsewhere, we read that "The Conference, sitting as the Biometric Society, adopted the Constitution as amended during

Saturday morning session -- page 3

The motion was made an seconded to call the new Society the "Biometric Society" and to add a small descriptive paragraph to the letterhead clarifying the objective of the Society.

<u>1. Scope of the Society</u>

<u>Kossack.</u> Suggested changing second sentence to insert "as members" after "welcomes".

<u>Norton.</u> "welcomes to membership".

<u>Geiringer.</u> Omit word "effective in third line.

<u>Tukey.</u> leave word "effective" in.

<u>Rashevsky.</u> remove word "effective".

<u>Hotelling.</u> pointed out that word "effective" had an affective meaning.

<u>Wehr.</u> suggested rearranging the words "Application, dissemination and development" to "application, development and dissemination". Wished to know the reason for putting "development" last.

<u>Tukey.</u> Development is secondary for our group.

<u>Belz.</u> Suggested alphabetical order and avoid ranking

<u>Wilson,</u> Leaving all three words out and ending sentence after "biological science".

<u>Bliss.</u> Avoid seeming to be inclusive with no very well defined function. Some jealousy in functions of this society.

<u>Rubin.</u> Adopt Bliss' recommendations by modify to read "The Biometric Society is an international society for the advancement of quantitative biological science through mathematical and statistical techniques".

<u>Belz.</u> Interchange words "dissemination and development".

<u>Rubin.</u> Objects to word "applying".

<u>Fisher.</u> Pointed out the wording is clear and should leave as stands.

Motion was made that the word "applying" be left out. Motion lost.

FIGURE 2.3
Minutes Saturday morning session, more heated and entertaining exchanges.

the morning session. The meeting adjourned at 1.20 P.M." The deed was done, the Society was born, and Bliss' "homework" was now graded!

With the inevitable spring in their steps and an excitement vibrating round the room, it was with purpose that the participants tried to concentrate on the afternoon scientific sessions. Nevertheless, from the Program of Figure 2.1, we observe that it was science, statistical science, biostatistical science, that bracketed the business sessions and that would constitute the foundations that drove the new Society forward.

The discussions on the Saturday morning (September 6, 1947) approved that anyone invited to Woods Hole including those unable to attend and all attendees were deemed to be Charter members. Nine days later (September 15, 1947) at the Second Council meeting held in Washington DC, this definition was expanded to include anyone who joined the Society before February 1, 1948.

THE BIOMETRIC SOCIETY—CONSTITUTION

1. *Scope of the Society*

The Biometric Society is an international society for the advancement of quantitative biological science through the development of quantitative theories and the application, development and dissemination of effective mathematical and statistical techniques. To this end the society welcomes to membership biologists, mathematicians, statisticians and others interested in applying similar techniques.

2. *Members*

The Society shall have one class of members. To become a member, a person must be proposed by two members of the Society and approved by the Council. The Council may delegate this authority. The members represent the highest authority of the Society. The Council shall consult them on any vital questions that may effect the policy of the Society as a whole, obtaining their decision by a mail vote.

3. *Officers*

The general officers of the Society shall be the President, the Secretary and the Treasurer. The regional officers shall be the Vice-Presidents, Regional Secretaries and Regional Treasurers. The officers shall be elected by the Council for one year.

The President shall act as chairman of the Council.

Each Vice-President shall represent a specified region. The regions shall be determined by the Council from time to time.

The Treasurer shall present financial statements to the Council and shall bring condensed statements to the attention of the members. The offices of Secretary and Treasurer may be combined.

No President or Vice-President shall serve more than two consecutive years.

The Council shall elect Regional Committees to serve under the chairmanship of each Vice-President and may elect Regional Secretaries and Regional Treasurers where that seems advisable.

The general and regional officers, acting as a nominating committee, shall submit to the Council a list of candidates for general and regional officers but members of the Council may vote for names not on this list. Regional officers and committeemen may be nominated directly by the members at regional business meetings.

4. *The Council*

The President, the Vice-Presidents, the Secretary, and the Treasurer shall be ex-officio members of the Council. There shall not be less than twelve not more than twenty ordinary members of the Council, who shall be elected with a view to representing the various geographic areas and fields of activity in which the Society has members. Ordinary members of the Council shall be elected for three years by a mail vote of the members, the terms of approximately one-third of them terminating each year.

The general and regional officers of the Society, acting as a nominating committee, shall submit to the members a list of names of twice the number of candidates necessary to fill the ordinary vacancies in the Council, but members may vote for names not on this list. No ordinary member of the Council shall serve more than two consecutive terms of three years each.

In matters of purely regional importance, the appropriate Regional Committee may act for the Council, but the Council may reverse its action.

5. *Activities*

Any activities which fall within the sphere of interest of the Society may be authorized by the Council, such as international and local scientific meetings and the issuance of publications reporting the activities of the Society or containing other matters of biometric interest.

The Society may affiliate itself with international organizations. With Council approval a Region may affiliate itself with regional or national organizations.

6. *Financial Organizations*

The dues for members shall be fixed by the Council. Requests and gifts may be received.

7. *Amendment of the Constitution*

Amendments to the Constitution must be approved by the Council, or in writing by at least five per cent of the members, and before becoming effective they must be ratified by a two-thirds majority of those voting in a mail vote taken among all members.

FIGURE 2.4
First constitution – approved September 6, 1947.

The First Council meeting was held that night, Saturday September 6, at Woods Hole. The first order of business was to elect Cox, John Burden Sanderson (Jack) Haldane, and Wilson as members (to those already approved that morning); Cox was duly fetched from her room to join the assembled crew. Then, Council elected Fisher as President, Hopkins as Treasurer and Bliss as Secretary. As Editor of *Biometrics*, Cox offered to publish the Conference Proceedings in the 1947 December issue; her offer was agreed upon (instead of an alternative proposal from *The American Naturalist*). The meeting adjourned at midnight; see Figure 2.5. Attendees at Woods Hole are shown in Figure 2.6.

The Second Council meeting was slated for September 15, 1947 in Washington, DC (where members were planning to attend the ISI Session). At this second meeting, with an eye to achieving balance across biometrical sub-fields, Adriano Buzzati-Traverso (Italy), Kenneth Stewart Cole (USA), Milislav Demerec (USA), Cyril Harold Goulden (Canada), Ivor J Johnson (USA), Joseph H Needham (UK), and George Gaylord Simpson (USA) were elected as additional members. The main order of business concerned the

Meeting of the Council, September 6, 1947 at 8:30 P.M.

Present: M. H. Belz, C. I. Bliss, R. C. Bose, R. A. Fisher, J. W. Hopkins, A. Linder, J. Monod, G. Teissier, G. Rasch, J. W. Tukey

Presiding: R. A. Fisher

... ...

The Council expressed a wish to meet in Washington during the meetings of the International Statistical Conferences after the officers of the A.S.A. had been consulted.

The meeting adjourned at 12:00 P.M.

R. A. Fisher 15 Sept 1947

FIGURE 2.5
Minutes - Council 1 - September 6, 1947, approved September 15, 1947.

Back Row:
J. N. Spuhler, Emily Jensen, John Watkins, John Rosenbaum, Mrs. Paul Bruyere, Lila F. Knudsen, Marion Sandomire, Harold Hotelling, Gertrude Cox, U. Chand, Hilda Geiringer, M. Lehr, C. P. Winsor, Alexander Weinstein, J. Lederberg, A. S. Householder, Mrs. Theodore Zucker, Theodore Zucker, K. W. Cooper, George Sacher, Lilian Elveback, Ruthedith Stigreaves, Unknown, R. L. Anderson.

Second Row:
Besse Day, Charles Cotterman, James Crow, F M. Wadley, James Rafferty, Paul Bruyere, R. Ruggles Gates, Oscar Kempthorne, Anotol Rapoport, N. Rshevsky, Alfonso Shimbel, Leslie F. Nims, Herman Rubin, Sidney Halperin, W. Ralph Singleton, L. N. Hazel, E. W. Lindstrom, J. W. Tukey, Frederick Mosteller, Churchill Eisenhart.

Front Row (Left to Right):
Carl Kossack, Otis A. Pope, A. E. Brandt, Mrs. H. W. Norton, Horace W. Norton, M. G. Neurdenburg, Maurice Belz, R. A. Fisher, G. Rasch, A. Linder, J. W. Hopkins, R. C. Bose, D. Catcheside, D. Nanda, W. G. Cochran, E. B. Wilson, E. J. deBeer, Charles Packard, Edwin Conklin, Chester Bliss.

FIGURE 2.6
Woods Hole attendees, September 5–6, 1947.

regional makeup of the Society (see Chapter 4). Another agenda item dealt with the subtitle for the Society's name and letterhead. The adopted resolution gave the wording "An International Society Devoted to the Mathematical and Statistical Aspects of Biology." The discussion on dues and related cost of *Biometrics* settled on $4.00 for dues but with a rebate of $2.00 for members who were already receiving the journal as members of the Biometric Section of ASA.

3

After Woods Hole

3.1 Overview

After the euphoria of its establishment at Woods Hole, there was now a Society to be nurtured and maintained.

The reports in *Science* in October 1947 and *The American Statistician* of the Society's formation at Woods Hole brought a slew of letters from those interested in joining the Society as Charter members or (later) simply as members.

One of the early preoccupations concerned its flagship journal *Biometrics* which represented the scientific backbone of the Society. The journal belonged to the American Statistical Association (ASA), and it was they who controlled policies and management questions. There was an additional complication that, due to its own overall operational losses in 1948, ASA assigned financial responsibility of its journals to the respective editors in 1949. By 1948, the Society wanted its own journal. Discussions with ASA about transferring *Biometrics* to the Society were initially quite cordial and fruitful; ASA President George Snedecor in 1948 even declared that the journal should "follow the Biometric Society rather than the [ASA]"; but a year later in December 1949 Snedecor suggested "[the action] was premature ... but look to the future." An effective transfer date of January 1950 was targeted. Committees were set up by both the ASA and the Society to determine details. However, as time moved on, frustrations surfaced with a looming apparent reluctance on part of the ASA leadership (but by no means all). Though the path was rocky, eventually ownership was transferred in October 1950. Another marker occurred in 1958 when it was decided to terminate a previous reciprocal block subscription to ASA members. The journal, now very much internationalized, was a scientific success. More details of these events, as well as descriptions of the editorial processes and journal content and related aspects, are presented in Chapter 7.

Apart from *Biometrics*, probably the biggest immediate concerns after Woods Hole were the growth in membership and the provision of sufficient funds to maintain the Society.

One source of funds was naturally the dues income from members. These were structured around the cost of publishing the journal *Biometrics*. Initially however, this income had to be supplemented by other sources. The Rockefeller

DOI: 10.1201/9781003285366-3

Foundation had provided funds to establish the Society in 1947. In March 1948, this Foundation awarded the Society additional funds totaling $7400 for the next three years to February 28, 1951 (up to $4000, $3000, and $1500, respectively) to be processed through Yale University since the Society had not yet established its tax-free status; these were to be used to employ a half-time executive assistant among other expenses.

However, growth in membership was the key. Chester Bliss was tireless and indefatigable, ever energetic, ever creative as he sought ways and avenues to grow the Society in both membership and its outreach and influence. No opportunity was missed, no matter how remote. For example, when ballots were sent to Council, each councilor was asked to write names and addresses of potential members on the back of the ballot. Members of related organizations, such as the ASA Biometrics Section, American National Red Cross, the Hawaiian Pineapple Company, biologists, and so on, were invited. Judging by the number of letters kept in the Archives, the net was cast wide and frequent. Bliss used every conceivable ploy to entice new members. If someone asked him about some detail in a publication, he would write back and ask "... In view of your interest in [whatever], I believe you would enjoy membership in the Biometric Society ...". Both Gertrude Cox and Bliss followed up with authors of journal submissions. Ronald Fisher was also part of these campaigns, e.g., " ... I have just heard that some biometrical activity is brewing [in Scandinavia] ... ." However, nothing can compare even remotely with Bliss' Grand World Tour (at no expense to the Society) when, during his sabbatical year (September 1961 to September 1962), he visited 30 different countries lecturing on biometry and the merits of Society membership; the archival cryptic details are four pages long listing off the 110 talks and even include the detail that "Day lost crossing International Data Line" and that he was "Bedridden with influenza" in Athens Greece – an extract is summarized in Table 3.1. Then, he would follow up after the visit/s to enquire about progress. Thus, it was that members joined, national groups formed some of which became regions.

Part of Bliss' mantra here included possibilities for locations of International Biometric Conferences (IBC) or Symposia in various areas of the world, hoping that such meetings would foster interest in the Society. These outreaches were often most successful, the shining example being the 1955 Symposium in Campinas, Brazil (see Chapter 9, Section 9.2.2). However, to Bliss' eternal dismay, he was never able to generate renewed interest in re-forming the Indian Region through the 1951 Indian Symposium, nor did his invoking of Fisher's interests in India upon Fisher's death (in 1962) bring any positive response. While these two undertakings were notable in their different ways, Bliss continued his efforts undeterred. Indeed, in the first two-three decades, among other issues such as coordinating with ISI meetings or other major conferences, the entire leadership spent a lot of time scrambling for potential sites and hosts for IBCs and symposia, as well as expending a lot of energy in more scrambling for funding support for them. In a reverse direction, Cyril Goulden accepted nomination for President (1958–1959) as he says "this will assist in the Ottawa Conference" (he was elected).

TABLE 3.1
Bliss' Grand World Tour September 1961 to September 1962

Country	Dates	Places Visited
1961		
USA	September	San Francisco; Honolulu
New Zealand	September	Auckland; Wellington; Christchurch
Australia	September	Melbourne; Canberra
	October	Canberra; Sydney; Brisbane; Hobart; Adelaide;
	November	Perth; Darwin
1962		
India	February	Lucknow
	March	Delhi; Agra
Iran	March	Tehran
Lebanon	March	Beirut
Jordan	March	Jerusalem
UAR	March	Cairo; Luxor; Alexandria
Turkey	March	Ankara; Istanbul
Greece	March	Athens; Thessalonika; Sindos
	April	Athens (again)
Austria	April	Vienna
Poland	April	Warsaw; Wroclaw; Krakow; Warsaw (again);
USSR	April	Moscow; Leningrad; Moscow (again)
Denmark	April to May	Copenhagen; Springforbi
Sweden	May	Stockholm; Uppsala
Norway	May	Oslo; Ås
Germany	May	Frankfurt
Switzerland	May	Zurich; Geneva
Italy	May	Pavia; Milan; Florence
France	May	Paris
Belgium	May to June	Brussels; Ghent; Liege
Netherlands	June	Amsterdam; Leiden; Utrecht; Wageningen; The Hague; Delft; Eindhoven
England	June to August	Cambridge; Bentley; London; Harpenden; Babraham; Bracknell
Scotland	August	Edinburgh
England	August	Cambridge; Harpenden (again)
Ireland	September	Dublin

Committees emerged as needs arose. Thus, the problems over finding IBC sites elicited the need for a Future Meetings Committee in 1962, first chaired by Bliss until 1970, followed by S Clifford Pearce (BR). As part of the ongoing constitutional revisions, in 1974, this became the Long Range Planning Committee, with the Secretary as its Chair and three other members appointed by the President. This Committee's primary role was to make recommendations about sites and hosts for future IBCs but also for smaller scale meetings. [In 2012, this Committee was renamed the Conference Advisory Committee.]

The core membership consisted of "regular" or "ordinary" members. Initially, potential members were identified as having a primary interest in designated subfields and had to be recommended by two members and approved by Council. The Council role was removed in 1974. In addition, there were associate members, student members, and sustaining members, which became "institutional" in 1983 or "corporate" members in 1998, and a "senior retiree" category. Policies governing these latter membership categories became part of the Council By-laws. The question of life members came up as early as 1952, and although a sliding scale based on age (age 62, 63, ...) was approved by Council in 1963, ultimately it did not seem to gain any traction.

A lot of attention revolved around the concept of Associate members. What constituted an Associate member took different forms over the years. Basically, though, there were two main directions. One direction involved those who received the journal (*Biometrics*) some other way, such as through membership of an affiliated association (as proposed by the Eastern North American Region (ENAR) in 1949) or as a collaborator of another member. Associate members had all the rights of membership except they could not vote nor hold office, and could not receive *Biometrics*; within a region, the number of Associates was not to exceed the number of ordinary members. The other direction, discussed considerably especially in the 1960s and 1970s, was motivated by a desire to enable potential members in Eastern Europe and other developing countries to join the Society and, in particular, to receive the journal. Many formats were discussed and discarded for various reasons. In a different direction, the Western North American Region (WNAR) proposed student membership in 1954, later identified as Student Associate members (in the 1998 revised constitution) who had the rights of associates except that students could also receive *Biometrics*; eligibility rules and tenure (initially at most three years, until 1983) varied, but basically those for regular membership pertained while being a full time student at an accredited school or institution. Associate and Student members enjoyed a reduced dues structure. A related but different category was the Senior Retiree member, defined in 2012 as someone who had been a member for at least ten years and who had retired from regular employment, also had reduced dues but retained the rights of a regular member. Fifty years earlier, first in 1956, a French Region proposal was considered, and then in 1960, Treasurer Allyn Kimball had asked about "emeritus" members; although never formally added as a category to the Constitution, some financial reports show reduced dues for retirees. During the 1960s, similar notions were posed for past officers, maybe as "emeritus" and/or non-dues for retiring officers. However, officers themselves rejected this, adding emphatically that officers should not receive any payment of any form for their service (a sentiment that was echoed a number of times across the years).

In the 1960s, the concept of Honorary membership emerged. Although Bliss and Cox (and Fisher effectively, had he still been alive) had been invited to be Honorary Life members by Council in 1963, British (BR) Regional

President John Skellam reported to Cox in 1968 that Frank Yates was retiring (but not retiring) and proposed that Yates too become an Honorary member. Cox asked Skellam to wait as a committee (chaired by David Finney) was working on proper procedures and rules to cover this category of membership. Cox also told Skellam that she found this to be "extra embarrassing" as she had been one of those so honored before the rules had been determined. The By-laws additions to allow for this contingency were subsequently approved by Council in 1969. Interestingly, though aware of this approval, the BR never did nominate Yates then for the first (1970) round (but he was so elected in 1971). Nominations for Honorary Life members had to include input from that member's region. These members retained the rights of regular members. During the time of these deliberations (mid-late 1960s), the concept of regions offering their own regional honorary fellowships was advanced, but it seems regions did not take up this option. When initially approved as part of the 1974 constitution revisions, a limit was set to be at most two per year; the 2012 constitutional revisions reduced this limit to no more than one new honorary member in a two-year period. In a different direction, in 1995, Council approved the concept of Fellow (as distinct from an "Honorary" Fellow), but this effort stalled and never materialized.

Ultimately, the various membership categories described thus far were outgrowths of the fundamental need to have effective members. However, from the outset, obtaining adequate funding to run the Society was also a foremost concern. In this connection, the idea of having sustaining members was first raised in January 1950. Bliss explored possibilities throughout 1950, explained to (now) President Arthur Linder (At-Large member from Switzerland) in December 1950 that it was important to "line up American firms before tackling overseas." A Council By-law to allow for such members was passed in February 1951. A committee was set up to facilitate the process. Letters to potential companies were duly sent. While some declined, many were enthusiastic, so much so that Bliss exclaimed that "this enterprise is proving very successful." By mid-1952, eight companies had become sustaining members[1]. Bliss declared: We can now issue a new directory in 1953. There is no slowing Bliss down; he continued to seek more such members, even offering to put a sustaining member representative on a committee. By 1953, outreach to European countries began (but not to the British Region who at the time opposed this type of member). Then in 1955 Bliss, referring to the "great turmoil over the Salk vaccine," suggested that (certain Salk-related) laboratories should be sustaining members; clearly, Bliss thought outside the box and did not miss a trick! This was the 1950s, and continued over the decades usually with encouragement from the regions. Dues for sustaining

[1]viz., Abbott Laboratories, Armour and Company (Research Division), Bristol Laboratories Inc., Collett-Week-Nibecker Inc., Eli Lilly and Company, Sharp and Dohme Incorporated, Smith Kline and French Companies, and Wyeth Institute of Applied Biochemistry.

25th February, 1958.

Dr. C.I. Bliss,
Drawer 1106,
New Haven 4,
Conn., U.S.A..

Dear Chester,

I thought you could not afford to miss this. Flourishing learned society going cheap! Own journal! Solvant!!

Yours sincerely,

M.J.R. Healy

FIGURE 3.1
Healy letter to Bliss February 25, 1958.

members were proportional to the member dues of the region in which the sustaining member was located.

That the founding fathers (and mother!) had established a well-structured entity was proven by events in the mid-1950s. The Society survived Cox's resignation as *Biometrics* Editor in December 1955, and even weathered the disruptions swirling around her successor's illness and need to step down within 18 months (see Chapter 7). Bliss wanted to resign in 1955 as Secretary-Treasurer; the Treasurer office had been joined to his Secretary role in 1951. Finding a suitable replacement was proving difficult. Eventually, after a letter (a masterpiece of tactfulness and persuasion) from British Regional President Finney, Michael J R Healy (BR), who importantly was fluent in English and French, agreed to serve as Secretary but not as Treasurer, effective from 1956. Meantime, Bliss continued as Treasurer for another year, until the appointment of Allyn Kimball (ENAR) was made to begin in February 1957. A potentially catastrophic period saw the reverse with the Society sailing through safely. Healy's letter to Bliss in 1958 says much; see Figure 3.1.

Invitations to Bliss to run as Vice-President during this time were declined, however; he wanted to finish his book though he said he might consider it in 1960. He did become President in 1962, poignantly so at the time of Fisher's death. On hearing of his (by the 1963 Council) election to Honorary fellow, Bliss in typical humbleness replies to Finney "Looking back at my brashness in 1947, it is hard to realize our good luck. To start our seventeenth year with 2253 paid-up members, well distributed over the world, augers well for the future." Thus, while choppy waters would also lie ahead at times, the Society was now a force – that "homework" had been completed indeed.

Many major issues dominated the 1960s, keeping the leadership busy. Fisher died (July 29, 1962); Healy as Organizing Chair of the 1963 Cambridge IBC re-crafted the Conference to be a celebration of Fisher's scientific career. This is described in Chapter 9, Section 9.1.5.

That 1963 IBC was also notable for the many initiatives adopted by Council. The Society Vice-President position was approved. An Awards Fund Committee was proposed and approved at the Cambridge 1963 Council meeting supported by $1500 per year starting June 1964 for five years for projects such as Sir Ronald Fisher memorial appeals, visiting lectureships, traveling lectureships, student scholarships, and the distribution of complete sets of *Biometrics* to libraries in need in developing countries. The first Chair was C R Rao (GInd). By 1969, no funds had been expended; the Committee had become "a mystery" to many officers. Unfortunately, the Committee had "never gotten into action" (Executive Committee Minutes August 18, 1970 and August 16, 1971) despite efforts from President Bertold Schneider encouraging Rao to re-activate the work of the Committee. President Peter Armitage also wrote to Rao in early 1972; this time Rao replied saying he "ha[d] only ever received one letter about the Committee and .... [did] not know either who the other members [were] or the terms of reference." It was not only a "mystery," the archival records are very confusing. With Rao's resignation in 1972 (because "Mahalanobis [had] expired"), the committee was resurrected. Finding members and a Chair proved problematical. Bernard Greenberg (ENAR), though willing to be a committee member only, served as the de facto Chair, until Joseph Gani (BR) became actual Chair in 1973. After a rocky start, the Committee work was underway. While, as might be not unexpected, there were some extraordinarily self-opportunistic applications for receipt of an award, the Committee held to the maxim that awards were primarily designed to enhance the Society and its members especially in geographical areas where the use of biometry might be otherwise inaccessible, i.e., to strengthen non-developed countries and not for renumeration of folk in developed countries. The Tokyo Council in 1982 extended the appointment term of Committee members from three to six years; members would represent three different continents, and in the last two years, the appointee would be Chair.

Finance committees had sometimes been set up by conference organizers, and such a committee apparently existed in earlier Council By-laws, but its membership and duties had never been specified, and so effectively it did not exist. Therefore, the Cambridge Council also approved Treasurer Marvin Kastenbaum's formal proposal of January 1962 that a Finance Committee be formed to advise the Treasurer. This Committee consisted of four members for four-year terms from four different regions appointed by the President with approval by Council, one rotating off each year; the Treasurer and Managing Editor (of *Biometrics*) would serve in an advisory capacity. The first Chair was

Lyle Calvin (WNAR). The Committee's role was to inspect and coordinate annual budget proposals from the Secretary, Treasurer, Editor and Managing Editor, and was to be responsible for annual audits of Society books (by qualified professional auditors, i.e., not by members). This Committee had to approve the budget prepared by the Treasurer, before it went to the Secretary for distribution to Council for final approval. During the 1960s, the importance that there be some financial reserves emerged. Indeed, Treasurer Henry Tucker triumphantly reported that for 1968 there was a "surplus of $76.99." Recommendations for dues increases would be channeled through this Committee; this aspect was vital especially in inflationary times (as in the early 1970s coupled with US dollar devaluations, and again in the 1980s) when struggling with securing the financial needs of the Society. The two devaluations in the early 1970s drew considerable attention; these would have an adverse impact on the two North American regions who had already dealt with higher dues than other regions paid, and the Society felt the time had come when this imbalance should perhaps be corrected with some form of dues equalization. By early 1970, there was a reserve, so that even the "bit of a loss" in 1972 did not cause too much undue consternation. However, by now, the struggles of the first decade for financial solvency no longer pertained. Nevertheless, it was 1994 before the Treasurer Stephen George could report that the Society was financially sound with an equity approximately equal to the then-current annual operating expense which was a minimum desirable level for professional societies.

Also, in the mid-1960s, ENAR engaged the leadership in an extensive discussion about the nomination process and regional makeup of Council. This was a protracted and at times quite animated discussion, even entertaining at times. New Secretary Henri Le Roy found this particularly difficult. Finney and Cox spent a lot of time trying to unravel the concerns; these are described more fully in Chapter 4, Section 4.2.1. One positive upshot of this occupation of energy was the revised constitution of 1974; see Chapter 10, Section 10.1. Another preoccupation of the 1960s, spearheaded by the British Region, concerned an apparent perceived failure of adequate communications between the Executive Committee and Council and the regions; more details can be found in Chapter 4, Section 4.2.2. Again, as often was the case when identified problems led to better solutions, the Society leadership had already set in motion a study of how best to manage overall operations; this resulted in establishing a Business Office to handle journal and society business matters (see Chapter 10, Section 10.2). Of course, communications work both ways. President Richard Cormack (1980–1981) opined that interesting things happened in the regions, but we, the Executive/Society, only knew what happened in our own region; he adopted Editor Peter Armitage's idea to put brief reports in *Biometrics*, and indeed this became a big part of the *Biometric Bulletin* with its launch in 1984.

There were, as expected, numerous letters seeking nominations for officers, Council, and committees. Then, there would be elections; not everyone was elected. After the 1971 election round, in a nice touch, Secretary Hanspeter Thöni sent letters to those who were not elected to Council in their own language. After all, the Society is an international society!

Secretaries were, always, the repository of all things societal. The President for 1974–1975 was C R Rao; on assuming this office, he wrote to Secretary Thöni in February 1974 asking that "As I am not in touch with the Society ... I have to depend on you for taking decisions." Rao continued that he wanted Thöni to tell him the what-how-who of situations and that he (Rao) would write and sign the letters. While Secretaries were mostly on top of things occasionally prompting officers about upcoming duties, Thöni found this difficult, but he was up to the task. Eventually however, after four efforts (three letters and verbally at the August 1974 Constanza IBC), he apparently finally convinced Rao he wanted to resign (by the previous December 1974), imploring Rao to "Please induce the election of a new Secretary as soon as possible." A new Editor was also needed; so (now 1975), Rao is asking Thöni "as I do not know the procedure ... will you do it?" On another occasion "since [I] don't know the officers, [will you do ...]?" – Thöni promptly mailed Rao another list of officers; and on it goes, with more examples asking Thöni too many questions and with Thöni politely replying as to what to do. The Archives revealed a picture that is just incredible! It is not clear what was behind this, though the Archives did tell us elsewhere on an entirely unrelated topic that clerical assistance from the region could be unreliable. However, Rao was proactive in the (unsuccessful) efforts to have an award honoring George Snedecor upon his/Snedecor's death (February 1974) as part of the Awards Fund Committee's functions.

The major push through the 1980s concerned President Pierre Dagnelie and Treasurer Jonas Ellenberg's "globalization" Manifesto wherein they worked most successfully at expanding the Society to areas not then adequately covered (see Chapter 6, Section 6.8, for details), as well as a concurrent effort to increase the sustaining membership portfolio. By the 1990s, the superinformation highway and the computer revolution (in its many facets) was a real force requiring concentrated attention. Therefore, in 1995, President Lynne Billard set up a Technology Committee (Chair, Adrian Bowman, BR). Committee members represented a broad range of the Society coming from all but one continent, some with considerable technical skills and some who barely had the opportunity to use a computer; since even the production of the journals would be impacted, journal editors were ex-officio members. The charge was to study the issues and make recommendations, all designed to keep the Society current and prepared to function in the decidedly new technological future. A website initiative begun in 1995 became reality in 1996; simultaneously, regions were encouraged to set up regional websites (though it must be said that some regions had already done so).

Before these enterprises involving the new electronic train, however, and not unrelated to the intrusion of the computer, the time had come when modernizing the business side of the Society became a necessity. Thus, in

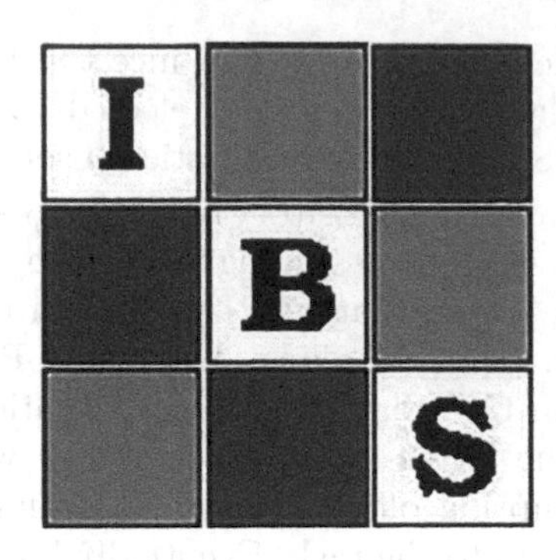

FIGURE 3.2
Society logo, created 1995.

1992, President Niels Keiding set up an ad hoc Committee on Business Management (with Treasurer Stephen George as Chair). This led ultimately to moving business operations to a professional organization management firm; see Chapter 10, Section 10.2.

Also, in the 1990s, a Society logo was created. This was designed by George and Billard, sitting in an Amsterdam restaurant, in 1995. The Executive Committee had met in Amsterdam simultaneous to overseeing aspects of the pending fiftieth celebration 1996 IBC with the Local Organizing Committee. George worked in cancer biostatistics and so recalled an elegant cancer center logo which featured a "broken" cell in a 3×3 matrix of small squares. This recollection led to the shaded 3×3 Latin square design in recognition of the experimental design work of founding President Fisher; incoming President Byron Morgan was asked to select the color (three shades of blue, not shown). Professional artists finished the product. See Figure 3.2. There had been an earlier unsuccessful attempt to solicit proposed logos from the membership, but this prior effort did not produce a suitable design. As an addendum, it is noted that the *Biometric Bulletin* Editor Joe Perry raised the question of a logo in 1989, but at the time his idea fell on stony ground. Initiated by Billard, trademarks for the logo and for *Biometrics* were eventually registered in 1999.

The so-called Club of Presidents was formed in 1994. This Club consisted of all former Presidents after their out-going Vice-Presidency, chaired by the second most recent President (the first Chair was Richard Tomassone). Its primary role was to act as a sounding board for officers who succeeded them should the need arise. Typically, they meet formally at IBCs. Their first substantive task was to be the Advisory Committee for the fiftieth anniversary celebration at the 1996 IBC, including being the Discussants for the special Anniversary Session at which former President and Editor Peter Armitage was the featured speaker (see Chapter 9, Section 9.1.18).

As an organization whose primary scientific focus was witnessed through its publications, the 1980s and 1990s saw more activity in launching new outlets. Always, the role of *Biometrics* as a newsletter continued until 1984 when the *Biometric Bulletin* began as the primary newsletter forum. Then, after years of perennial complaints about *Biometrics* being too mathematical which expanded to include complaints that the journal was too medically and epidemiologically oriented, a new interdisciplinary journal with a focus on "... the development and use of statistical methods in the agricultural sciences, the biological sciences including biotechnology, and the environmental sciences including sciences dealing with natural resources" was launched, jointly with ASA, in 1996 as the *Journal of Agricultural Biological and Environmental Sciences (JABES)*. These publications, along with other Society publications and those published by regions, are discussed in detail in Chapter 8.

3.2 Regions and Groups

Cutting across these decades was the formation of regions and national groups. A distinctive feature of the Society was that its international core was defined by its geographically situated regions and national groups, forming the spokes (so to speak) of the international umbrella.

At the Second Council meeting held on September 15, 1947, the major item of business was to set up a skeleton regional organization. The first region formed was the Eastern North American Region (ENAR) in December 29, 1947. The British Region (BR) formed in April 29, 1948 in London. The Western North American Region (WNAR) became official on November 27, 1948 during a concurrent Institute of Mathematical Statistics (IMS) meeting in Seattle. The Indian Region (IR) was organized on January 5, 1949, but soon thereafter because of a dispute over dues, the region disbanded, eventually became a National Group (GInd) in January 1953, and then reformed as a region in 1989.

The Second Council also identified Scandinavia, Benelux countries, Australia, France, and Russia to be regions, even designating contacts to explore respective regional formations. The Australasian Region (AR, consisting of Australia and New Zealand) was quickly established in 1948. In Europe, an initiative (begun in February 1948) by Italian member Adriano Bussati-Traverso after contacts with Frenchman Georges Teissier, conceived of forming a region involving France, Italy, Switzerland, and Benelux. In the end, France went its own way forming the Région Française (RF, French Region) in March 15, 1949 in Paris, and simultaneously the Société Française de Biométrie, with members of this latter society to be members of the French Region as well, so as to satisfy a 1901 French law. The French paradigm also pertained to members in Belgium when they formed a Belgian Region (RBe) in 1952 with members simultaneously part of the Société Adolphe Quetelet. Progress toward a region in Italy was somewhat slow although enthusiastic, with the formal Council approval in 1953, as the eighth region (RItl).

The Netherlands, part of the Benelux countries, became its own region (Afdeling Netherland, ANed) in 1970 having been a national group since 1949. It was not until 1982 that Scandinavian members became the Nordic Region (NR), which came as a result of merging national groups in Denmark, Norway, and Sweden, along with At-Large members in Iceland and Finland.

Along the way, the Society continued to expand and saw national groups become established in many geographical areas of the world whenever the membership count reached about ten. Most of these national groups became regions as their numbers increased beyond 50 (usually, though in earlier years this number was near or at 35). The first was the German Region (DR), the Society's ninth region, which became official in January 1955. The Brazil Region (RBras) quickly followed in January 1956. The Swiss Group expanded to include Austrian members and became the Austro-Swiss Region (Region Österreich-Schweiz, ROeS) in January 1962. Japanese members had first approved that they become a region back in August 1953; however, it was 1979 before it became the Japanese Region (JR). The Hungarian Group became the Hungarian Region (HR) in October 1988 but was dissolved in 2003 to return to group status. In 1992, the Spanish Region (REsp) became the Society's final region to be formed in its first fifty years.

Efforts by Bliss and later by successive Society executives to establish groups in Eastern European countries proved difficult, primarily because of currency and political issues. The hoped-for Russian region at the Second Council in 1947 did not materialize, but not from lack of effort and desire (Bliss, in particular, was tirelessly active trying to establish a sizable membership not just in Russia but also in other Communist countries). Potential members were hampered by difficulties in exchanging currency, dues, and the relative impoverished level of salaries as inhibiting factors. Considerable time was expended by the Society leadership seeking ways to resolve attendant issues. It was not until the end of the Cold War that real progress occurred. Several regions of the more affluent Western countries were also involved in sponsoring members or libraries etc., not just in the Eastern European countries but also in other regions such as African nations where interest in biometry was clearly high. Chapter 6, Section 6.8, provides more details of this growth and explores ways taken by the Society to enhance membership casting a wide net.

Chapters 4–6 provide more complete details of these regions and national groups, their formations, and highlighted events.

This account obscures a lot of interesting activities within and between regions and groups over the years. One intriguing example pertains to Germany. After the Berlin Wall was erected, difficulties naturally arose because there were members on both sides. To quote Bliss (1965), "Within this one Region we find the kinds of political complications which dominate international relations today." However, the Society did untangle the complications within the German Region. Eventually, in 1969, a new East German Group was formed, but by 1971, it became the German Democratic

Republic Region (RGDR). This story ended with the reunification of the two regions in 1991. In quite a different direction, other major efforts involved ENAR and British Regions "complaints" to hear Cox say it; these too were resolved in ways beneficial to the Society; see Chapter 4.

3.3 Meetings

The journals as well as the international and regional meetings were the prime venues for the dissemination of scientific methodologies. Apart from the obvious scientific value of IBCs, committees and Council would meet during an IBC. As one President wrote "long distance written debates and discussions of issues leave a bit to be desired." Another asserted that "it is easier to resolve [an issue] by a meeting rather than by correspondence" and another "that treating items by correspondence makes the matter the more difficult." For the participants, "fraternisation is the chief aspect of conferences." And on it went. Indeed, the Society was founded at the first IBC held at Woods Hole.

In the early decades, the timing and location were deliberately chosen to precede or follow an ISI meeting or some other major scientific meeting, often an international genetics conference. Nevertheless, as indicated in Chapter 9, preparations for each successor IBC were sometimes fraught with a desperation as to when and where the next IBC should be held. Thus, Conferences were held at irregular intervals, depending on the success or otherwise of garnering a suitable host. Council at the Sydney 1967 IBC suggested that there must be a maximum of four years between Conferences. Then in 1982, Council declared that future IBCs should be held every two years and in off-ISI years; this was a momentous decision and completely underscored the fact that the Society was now an independent entity, strong in its programs, and no longer felt the need to tie itself to the ISI programs.

Details of the complement of IBCs are given in Chapter 9, Section 9.1. In the early years, the Society was dependent on funding support from agencies such as the International Union of Biological Sciences (IUBS) though IUBS had limits on the frequency of its support. This meant that Symposia would be held instead; see Chapter 9, Section 9.2. The scientific programs became a microcosm of prevailing new developments. Over time, the programs separated out into invited sessions and contributed sessions in a variety of formats; eventually, there were so many participants, the one-at-a-time session yielded (in 1970) to parallel sessions. Furthermore, as a reflection of the international persona of the Society, these meetings were held in interesting locations throughout the world, hosted by the local region.

Some funds were provided by the Society for basic conference expenses, though ultimately some hosts declined to accept funds, while some hosts reimbursed funds or loans available to start the organizational processes. The question of whether or not invited speakers should receive some travel support was a perennial topic over the years, with differing conclusions at different times. There was a general reluctance, however, to increase conference registrations to allow for such support, especially as this would adversely impact participants from less affluent countries. After much discussion by the 1988 Namur Council, the Society declared that it must protect itself financially (from losses). Then, in 1990, the Society recognized that in general (major exceptions being the 1967 Sydney IBC and the 1986 Seattle IBC), the Society "consistently lost money on the IBCs," whereas many regions and organizations used meetings to generate profits thus reducing the impact on overall dues paid to the Society. Therefore, the Society proposed that "invited speakers [could] no longer be paid to present..." Also, since some regions, most notably ENAR, had found that offering short courses had proven profitable and popular, the Society also proposed that short courses be sponsored at future IBCs. These short courses became a fixture and later were the responsibility of the Education Committee. The 1990 Budapest Council approved that the registration fee and up to a maximum of $200 could be offered to at most 20 invited speakers for future IBCs. It was never permissible, however, to give honorariums to speakers (as this violated the constitution in that no member could be paid for services to the Society).

In the mid-1980s, the Long Range Planning Committee appointed a committee to prepare a manual for future IBCs (and Symposia); Gerald van Belle (Local Chair of the 1986 Seattle IBC) was chair. The final 69-page Manual, submitted to Council in 1990, covered all aspects of running a conference; it was updated periodically, and by 2012 was 160 pages. A parallel Manual for IBC Program Committees was also prepared; in 2006, this was 59 pages. The Society clearly took the role of successful conferences very seriously.

Council always meet during an IBC, as did committees. It was here that Council meetings at the 1963 Cambridge and 1982 Toulouse IBCs were particularly important for the many major policy initiatives that emanated from those respective Councils.

3.4 International Affiliations and Associations

National and international affiliations were sought from the outset. Regions could affiliate with their local national organizations but not with international organizations, while the Society could affiliate with international organizations.

Council at its second meeting September 15, 1947 authorized Fisher to approach the International Council of Scientific Unions (ICSU) in regard to affiliation with the Society. Thence, a proposal for an International Union of Biometry (IUB), submitted in November 1947, was seen as a way of securing this affiliation. The hope was that IUB would organize international conferences. Not embraced by Council, thoughts moved to the International

to this idea, ~~that~~ then it would be perfectly fatal to leave the interests of statistical methods in biology to a couple of biologists who may well have been occupied for most of their lives in resisting the use of statistical methods.

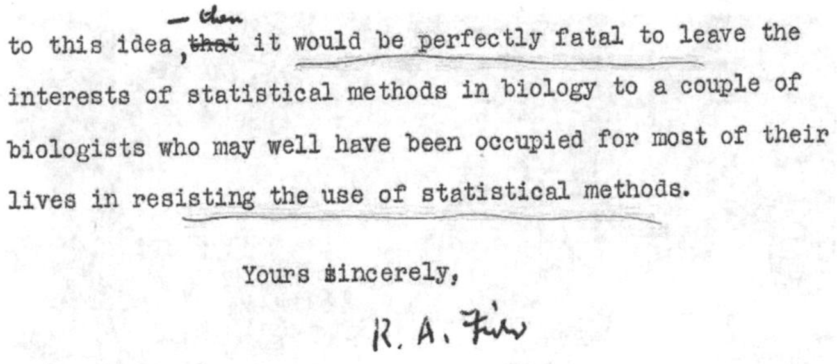

Yours sincerely,

R. A. Fisher

FIGURE 3.3
Fisher to Bliss, June 4, 1948.

Union of Biological Sciences (IUBS). Even so, hopes that the Society would be "fairly treated" were deemed "a little fantastic." Fisher felt "that it would be perfectly fatal to leave the interests of statistical methods in biology to a couple of biologists who may well have been occupied for most of their lives in resisting the use of statistical analysis." See Figure 3.3.

While waiting for the ICSU report on whether or not to re-activate this IUBS as one of its members, the Society continued its efforts regarding a proposed section on biometry, including the possibility that the Society be "a source of biometrical advice in the international field." In November 1948, IUBS Secretary-General P E L Vaysseière reported that the Society could be the Section on Biometry of IUBS and that this could serve as a "stepping-stone toward recognition of biometry by ICSU." Formal approval came in 1952. The Society's Secretary served as the Secretariat of this IUBS Section. That the Society had this function was evident by being featured at the bottom of Society letterhead (see, e.g., Figure 7.4 in Chapter 7). Reports on the Section on Biometry were required before each General Assembly; and IUBS felt it was "essential" to be represented at its General Assembly even reimbursing travel expenses for one representative. The IUBS was related to United Nations Educational, Scientific and Cultural Organization (UNESCO) but for administrative reasons direct affiliation was not possible. Funding support for conference proceedings would often come from either UNESCO or IUBS. The Archives include regular reports to IUBS; these tended to be summaries of recent IBCs and attendances, and updates on memberships and journals. There are also many letters regarding possible representatives to IUBS General Assemblies; as Bliss once said, being in attendance is important as we did not want to be forgotten. Figure 3.4 shows Society representatives André Vessereau (Paris, RF) and Luigi Luca Cavalli-Sforza (Milan, RItl) at the Rome 1955 General Assembly.

The ISI President Stuart Rice invited the Society to apply for formal affiliation, in 1948; the 1947 Council had already authorized this action. This was officially approved in March 1949, and there would be "an exchange of representatives between the two organizations." The Society's Secretary was this representative (at least up to 1963). Immediately, both organizations

FIGURE 3.4
Vessereau and Cavalli-Sforza at Rome 1955 IUBS General Assembly.

invited each other to their respective meetings scheduled in Switzerland in September 1949. In the early decades, that each society organized a scientific session in the other's major meeting was a regular occurrence. That said, over the years, this joint representation waxed and waned, so much so that with time it seemed that the Society's role as a session organizer at an ISI Session did not always eventuate. The ties were renewed in 1994.

The year 1949 also saw initiatives to become affiliated with the World Health Organization (WHO). This was approved by WHO in February 1954. As for IUBS General Assemblies, there are letters seeking suitable folk to represent the Society at the various WHO meetings.

Affiliation with ASA could only be effected through ENAR; this was approved in 1949. Earlier, ENAR had also affiliated with AAAS with an early 1948 approval.

In a different direction, occasionally over the years, outside entities would seek advice and/or input on some pressing (to them) issue. For example, the Food and Agricultural Organization of United Nations (FAO) requested that the Society provide an observer to "World Consultation on Forest Genetics and Tree Improvement" in Stockholm August 1963; this followed on the heels of a request that the Society be represented at the World Food Congress in Washington DC in June 1963. Bliss embraced these opportunities to extend the reach and influence of the Society; so letters would solicit potential members to act as the Society representative. Likewise, though different in precise nature, the Society was active in promotion of a Liaison Committee on Statistical Ecology in coordination with ISI, International Association for Ecology (INTECOL) and the field of statistical ecology. Requests to the Awards Fund Committee for financial support were not recommended by Finance Committee, despite the fact that members John Skellam (BR) and Douglas Chapman (WNAR) were appointed as Society representatives to this Liaison Committee.

4

Regional Structure and Early Regions

4.1 Introduction to Regions and National Groups

One unique characteristic of the Society was its regional and national group structure. From its beginnings, the Society was not to be a collection of independent national societies but rather an integrated whole consisting of autonomous but interdependent regions with some of the (international) Society functions delegated to the regional subdivisions, constituting umbrella spokes so-to-speak of the overall international society, yet each with its own identity. The regions and national groups were independent but nevertheless were part of the Society as a whole. They had their own governance (as prescribed by the Society Constitution), and they conducted their own scientific meetings (conferences or symposia) on a regular basis (again as prescribed by the Constitution). Geographical areas with larger membership numbers became regions, while those with smaller membership numbers were organized into national groups. It is this division of the world into separate regions but still linked under the international umbrella that distinguished the Society from other international associations.

At the Second Council meeting in Washington, DC on September 15, 1947, a skeleton regional organization was set up starting with four regions, namely, British Region, Indian Region, Western American Region, and Eastern American Region (interestingly recorded in the Minutes in this order); these latter two regions added the word "North" to their names when formed. The first region to form was the Eastern North American Region in December 1947, arising from the fact that the Biometrics Section of ASA had its annual meeting then. Likewise, the British Region formed in May 1948, and the Western North American Region evolved from the parallel meeting of the Institute of Mathematical Statistics (IMS) in Seattle in November 1948. Efforts to form an Indian Region began in 1949. Regions were also proposed by the Second Council for Scandinavia, Benelux countries, Australia, France, and Russia, with identified contacts to explore the respective regional formations. As described below, some of these took place reasonably quickly, others took much longer.

The Society also established national groups, where the number of members did not reach the requisite number of members to be a region per se (usually fifty), though there where some areas that had more than

DOI: 10.1201/9781003285366-4

fifty members but wanted to remain as a national group. The formation of some regions, or national groups, was reasonably straightforward, while the establishment of other regions was fraught with real and/or perceived political considerations. We look at each region and national group and its formation in detail in Chapters 4–6, presented in chronological order of its actual formation when approved by Council (i.e., not necessarily when negotiations started in pursuit of that formation). Members outside an organized region or national group were considered to be members in an "At-Large" region.

The legislative body governing the Society was the elected Council, while that for a region was the elected Regional Committee. As per the By-laws, regions had Regional Advisory Boards which established the slates for the election of the region's officers and regional committee members. At the beginning, all regional appointments and decisions had to be ratified/approved by Council. This included for example election results for regional Vice-President and also Regional Committee members themselves. Council at its 1963 meeting during the Cambridge International Biometrics Conference (IBC) decided that it was no longer necessary for Council to approve regional officers except for extraordinary cases. This change was officially approved in the 1974 ratification of the revised constitution. Over time, other functions were transferred to be solely a regional or national group decision. It was Council who was to keep its regional members informed of Society matters, and regional officers who informed the Society officers through their Council members; this two-way information traffic flow did not always function as well as it should.

Until 1953, the regional leadership that was later called "Regional Presidents" were Society "Vice-Presidents" and served on Council. This change in title was proposed in 1953, becoming effective in January 1954. The Regional Presidents were, however, still ex-officio members of Council until 1974.

Initially, the Society collected dues with varying, but Council approved portions handed back to the regions for local regional expenses. Eventually, beginning with the 1955 dues, except for the Eastern and Western North American Regions (ENAR-WNAR), this collection was handed over to regions/ groups so that instead of each member paying bank transfer charges, this banking cost was limited to the one transfer from the region or national group. The regions/groups retained their respective portions and forwarded over to the Treasurer the remaining totals collected for the running of the Society (a large part of which was to cover the costs of the journal *Biometrics*). Each region/group was required to provide a financial statement to the Treasurer each year. There were regular annual letters from regional treasurers with lists of those members who had not paid their dues; this information was critical for the Society so as to know to whom *Biometrics* should, or should not, be sent.

Dues were structured around the cost of the journal *Biometrics*. There were two small additional amounts, one to cover the costs of running the

Society (the international portion) and one to cover the costs of running the region (the regional portion, with suitable adjustment for a national group); all components were fixed and approved by Council. However, after the 1974 revisions of Council By-laws, regions and national groups could fix their own local dues without Council approval, though general dues (the international and the journal portions) still needed Council approval. In their first year only of formation, some national groups were able to retain the international portion to assist in organizing the national group. The role played by *Biometrics* was crucial, however. Without paid-up dues and without the resultant membership lists being sent to the Treasurer, *Biometrics* could not be mailed out.

Thus, it was especially important for regional treasurers and national secretaries to be diligent in collecting the annual dues and forwarding them to the international office in a timely manner. The Archives sport endless letters and communications dealing with this issue and its associated vexing time-lines. Sometimes it was a regional treasurer who appeared to be somewhat lax (or it may have been a regional secretary who had not sent out relevant dues renewal notices), sometimes it was a recalcitrant member, sometimes a member would write indignantly about not yet receiving the journal. Sometimes there were reports, usually from/to the Treasurer, describing delinquency rates from certain regions. Also, regional secretaries were responsible for distributing information and ballots where necessary to members, likewise national secretaries. For the purposes of dues collection, the Treasurer served as treasurer for the At-Large region. In short, the regional officers were key to the effective functioning of the Society.

Up to 1949, there was no lower limit to the size of a region, though all regions formed until then had thirty-five or more members. Soon thereafter (1952), the limit became a working fifty number. National groups were formed once there were ten or more members. [After 2012, all national groups were called regions.]

As a whole over the years, there were incidents within a region that dominated the attention of the leadership of the Society at the time. Though initially regional issues, some of these generated ideas for future consideration for the international society. These events, some comical, some painful, some just a product of what Gertrude Cox defined as regions who were sometimes exercising national proclivities instead of international bents, will be included under the respective regional sections below. As an aside, Cox shone through as a leader with a truly international perspective in her letters, but she was also very pragmatic when opining about events in some regions, most especially her own Eastern North American Regional colleagues in the mid-late 1960s.

In Chapters 4–6, we look at the formation of these regions and national groups, as revealed by the (sometimes quite colorful, sometimes less so) Archives rather than the sanitized Reports to Council or the published records in *Biometrics* prior to 1984 and in *Biometric Bulletin* from 1984. This chapter, Section 4.2, describes the development and defining events for those regions formed within the first year of the Society's existence. Then, Chapter 5 looks at the formation of regions over the next ten years. Finally, Chapter 6 considers regions which emerged in subsequent years; this includes national groups and networks. Though the prime focus is on the first fifty years[1] of the Society (through to 1997), a brief look beyond 1997 is included in Chapter 11. For the record, a listing of regional presidents and national secretaries is provided in Chapter 12, Section 12.2. The standard abbreviations for these regions and national groups are listed in Chapter 12, Table 12.1.

4.2 The First Year – Early Regions

4.2.1 Eastern North American Region (ENAR)

As described in Chapter 2, the Society was formed at Woods Hole in September 1947. In those days, the American Statistical Association (ASA) held its annual meeting in December. A key component of the Society's formation was the role played by the ASA's Biometrics Section.

Thus, it was that on December 27, 1947 in Chicago at an American Association for the Advancement of Science (AAAS) meeting, members called a business meeting (jointly with the Biometrics Section of the ASA) to establish the Eastern North American Region (ENAR) of the Society, setting out proposed By-laws and a slate of officers. A resolution was unanimously endorsed acknowledging the key role and support of ASA and Biometrics Section members and an express hope that ENAR and ASA would "establish and maintain the most cordial relations" with each other. A follow-up meeting was held December 29, 1947 in New York City, during the annual ASA meeting, to complete the actions begun in Chicago. By-laws were settled (still to be approved by Council) and officers were elected. Charles Paine Winsor (see Figure 4.1) was elected the inaugural Vice-President. Winsor, most noted for the winsorization method, was a biostatistician and engineer at the School of Hygiene and Public Health at Johns Hopkins University.

A Regional Advisory Board was also established at the December 1947 meetings, designed to advise the Regional Committee on all matters such as policy, nominations to committees, nominees for regional elections, interests of regional members, and so forth. At the beginning, the Regional Committee was a part of this advisory committee; later it became a separate entity but with the Chair of the Regional Advisory Committee being a member of the Regional Committee. Initially, a Nominating Committee prepared slates and

[1]In the early decades, the Archives contained quite a few records of most regions and national groups. However, unfortunately, as a general comment, the history of the development in later years is quite uneven because not all regions/groups have been as diligent as others at turning over their archival records to the international arm of the Society. Therefore, in that sense, the presentation in these chapters is perforce somewhat incomplete and quite varied in this coverage as it is limited to the records available.

FIGURE 4.1
Charles Paine Winsor (1895–1951).

recommendations for the election of officers until this role was taken over by the Regional Advisory Board. Election of the regional President-elect began in 1959 so as to facilitate continuity of the office of regional president; this regional president-elect became regional president the following year. This obviated the need for a regional president to serve two years (the first to learn, and the second to act in effect), and by doing so allowed for the honor and duties of being regional president to be shared by more members. The revised Regional By-laws in 1963 established that the regional President-elect would be responsible for programs for meetings that occurred during his Presidency year; he would appoint program committees for ENAR meetings and liaison with committees participating in joint meetings.

The ASA's Biometrics Section and ENAR became affiliated societies immediately upon formation. An ENAR representative was on ASA's Council, and the Regional Committee had an ASA representative. Good relations remain today. This included the fact that, then as now, there were to be joint meetings with ENAR and the Biometrics Section. When the ASA decided to move its annual meetings to early Fall for 1954 on a trial basis, the IMS approached ENAR in early January 1954 on this question. Chester Bliss reported back that probably ENAR would move too since the "influencers" in ASA were also those for ENAR; and so they did. These annual meetings slowly shifted from late December to the (northern) fall and then to the summer. Around the same time, by 1954, ENAR (still jointly with the Biometrics Section) began Spring meetings though the annual meetings at the time of the ASA meetings were maintained. Indeed, ENAR and the Western North American Region (WNAR) became two of the partners with ASA in the annual Joint Statistical Meetings (JSM). Prime responsibility for the organizational aspects of the Spring meetings lay with ENAR, while ASA assumed that role for the JSM summer meetings.

Because of the intertwining of ENAR and ASA's Biometics Section, both were heavily involved in the negotiations when the ownership of the journal *Biometrics* was transferred from ASA to the Society (described in detail elsewhere; see Chapter 7, Section 7.2). Indeed, the back-and-forth between the Society and ASA included discussions and concerns about the future of the Biometrics Section of the ASA vis-à-vis ENAR. Assurances were given to ASA that this would not lead to the demise of the Biometrics Section. On the contrary, both entities continue to exist and function independently, even though at times over the years officers of one were also officers of the other with both societies continuing to cooperate on the issues of the day. For example, the ENAR, WNAR, and Biometrics Section Chairs on the JSM Program Committee still coordinate all three roles today.

In early 1954, then regional secretary-treasurer Walter Federer reported an "unexpected windfall – unexpected to me at least" ($228.31) from the 1953 regional meeting in Washington, DC. Some of these monies helped support the free distribution of a "Career Opportunities in Modern Biology" booklet. The idea that meetings could be revenue generators was clearly a new concept, and while it was not overtly applied in subsequent years, budgets for meetings were now set at least not to lose money. By the mid-1980s, ENAR enjoyed a healthy surplus, so much so that one member – surely in jest – suggested that the next regional meeting should be held on a cruise boat. This activated the then-leadership to establish guidelines to regulate how any future budget excess should be spent; all guidelines focused on promoting the health and growth of the Society in general and ENAR in particular.

The first initiative was to introduce Student Travel Awards toward expenses to attend the Spring regional meetings to present their work, starting from 1984. This became a hugely competitive and highly effective tool in enhancing the Society but also the students' careers, still strongly in force today. Soon thereafter in 1988, the "best" such paper was named the "Van Ryzin Award," in honor of regional president John Van Ryzin who died in office (in 1987). Other initiatives were to introduce short courses immediately preceding the Spring meetings (but these made money from the outset), and to support libraries in developing countries. By the 1990s, ENAR was sending support ($5000 per year over 1991-95) to a range of developing countries such as Romania, China, Cuba, Argentina, Brazil, Chile, South Africa, Poland, Lithuania, India, and Thailand.

Those portions of the United States and Canada east of the 104° West Longitude line constituted ENAR; those to the west formed WNAR. In August 1962, there was a proposal from ENAR to merge ENAR and WNAR into one region with three sections to facilitate coordination between the regions. Council discussed this in August 1962 and wondered if there should instead be three different regions. At WNAR's August 1962 Business Meeting, there was a long discussion about this proposed merger. This was also discussed at

ENAR's September meeting; each region sent a representative to the other's meeting. This proposal was opposed by WNAR with then-President Bliss's backing. There was subsequently no further discussion of a merger, although the basic problems of regional coordination that had led to the merger proposal still remained.

In the mid-late 1960s, a major concern that occupied a lot of time from the ENAR leadership (primarily spearheaded by then-regional presidents Ralph A Bradley and Richard L (Dick) Anderson) revolved around the nomination process for Council and ENAR representation on that Council. The first indication of this issue was in a then-ENAR-President Bradley letter of May 1965 to Secretary Henri Le Roy about a perceived under-representation of ENAR on Council. There were endless letters between ENAR and Secretary Le Roy and President David Finney in particular with detailed criticisms drawing upon ENAR's interpretation of the Constitution and By-laws governing Council nominations and composition. Finney was masterful in his patience and tact and so very careful about it all in his dealings with the ENAR folk. He devoted a lot of time trying to ascertain just how and why ENAR was, in his eyes, misinterpreting the By-laws here, whereas ENAR was adamant that the Society officers were not following said By-laws in regard to the Council nomination process. Eventually, Finney determined ENAR was using the old Constitution and By-laws of 1953 rather than the updated revised version of 1962, a displacement of source that Finney found quite amusing. Later, in December 1965, in reply to Anderson, Finney is "glad to discover that some of our difficulties disappear when we all refer to the same set of rules," though Finney concluded with "ambiguities [still] exist ... ."

However, the consternation continued through Luigi Cavalli-Sforza's term as President (1966 and 1967), though Finney as out-going Vice-President in 1966 and Cox as in-coming Vice-President in 1967 continued to be the frontline spokesmen for the Society in the exchanges with ENAR, since Cavalli-Sforza would be largely out-of-contact while in "Africa on a scientific expedition." The discovery of ENAR's use of an outdated Constitution settled one question, but further complaints about inadequate/adequate representation continued. Finally, when Cox became President in 1968, she acknowledged her uncertainty about what ENAR really wanted; by January 1970, she wrote in exasperation (to now-Secretary Hanspeter Thöni) saying she hoped that she could find clarification to "stop all these long letters of complaint on Council elections" from ENAR. As was Finney, Cox also was tactful and courteous throughout the ordeal.

To his credit, Le Roy tried to defend ENAR though he did say in a May 1967 letter to Cox that he thought the process (he had followed) was "absolutely legal"; but Le Roy also found it all a hard grind despite the fact that Finney supported and encouraged Le Roy strongly, as did Cox. Cox (June 1967) drawing upon her extensive international experience (she had been an officer in the International Statistical Institute (ISI)), tried to console Le Roy that she understood well the complications of working with people from various countries and in reference to the ENAR troubles in particular thought it was "difficult for U.S. Statisticians to be humble." (Here, as elsewhere, Cox pulled no punches!) At one point (1967), Cavalli-Sforza found it prudent to assure the British Region that there would be "no hasty action" on the ENAR proposals. Civility did prevail throughout however, even to the extent that Anderson was invited to attend the 1967 Council meeting held during the Sydney IBC as an observer with a vote!

Council membership was limited to two times the number of regions (at the time, twelve regions, actually eleven but the non-regions counted as a single region for these purposes) but with the provision that all regions be represented and that each region have at least one member on Council. New regions were being formed, so that larger regions such as ENAR could see themselves losing representation relatively. In an exchange of letters between Anderson and Finney in December 1965, Finney agreed with some of ENAR's points but not with others, adding that the processes being applied were not necessarily undemocratic, and suggested that ENAR might want to propose a set of revisions for Council consideration. Indeed, the next year 1966, now ENAR President Anderson did set up its own ad hoc committee (with Herb David as Chair) to look at the Constitution which report was then sent to Society officers. The next several years saw lots of communications and debates on the issues. In some sense, this issue simply reflected the truism that, not for the first nor last time, situations can arise that reveal holes that necessitated amendments in the Society's Constitution and By-laws. While the dissension was long and protracted, a positive upshot was that in 1970 President Berthold Schneider set up a Constitution Revision Committee (with members Henry Tucker as Chair, Anderson and Finney) to study the issues carefully; a revised Constitution was sent to members on May 1, 1974 for approval; see Chapter 10, Section 10.1.

First, the region was managed by the regional officers. Discussions began in 1959 to transfer regional clerical functions to an outside agency. Subsequently, in 1965, Council approved ENAR's request that the management of the region's financial and business aspects be taken over by the ASA Business Office effective from 1966; the By-laws were changed to allow for the two offices to revert back to the combined Secretary-Treasurer structure that prevailed until 1961. With the establishment of the international Society Business Office in Washington in 1979, ENAR's management was transferred back to the Society Business Office. Then in 1992, ENAR management was moved to a professional organizational management firm.

From the beginning, voting for officers occurred at the annual meetings until 1956, when mail balloting came into effect.

A sense of humor would show through on occasions, as when regional Secretary-Treasurer Arthur M Dutten promised Bliss (April 8, 1954) a report "of my preliminary activities and *mistakes*" (emphasis added), or Regional President Mike Free declared that "not much was accomplished" though "we had a good time."

4.2.2 British Region (BR)/British-Baltic (BIR)

With its twenty charter members (defined as those at Woods Hole in September 1947 later adjusted to include those who joined by February 1, 1948, see Chapter 2), an organizational meeting was held on January 21, 1948. A provisional committee, set up at this meeting to bring the Region into being, consisted of John William Trewan (designated Vice-President), Kenneth Mather (designated regional Secretary-Treasurer), Ronald Aylmer Fisher (the Society President), John Burden Sanderson (Jack) Haldane (Council member), Edgar Charles Fieller and E C Woods, a distinguished group indeed. Trevan (Figure 4.2) was a medical researcher using statistics in biological assays among them the median lethal dose test. Thus, the inaugural meeting of the British Region (BR) was held at University College in London on April 29, 1948. The agenda included a report on the present state of the Region and a copy of draft rules (regional By-laws). Fisher gave an extraordinarily brilliant address (see Fisher, 1948; republished in 1964 in the special Fisher anniversary volume for "obvious reasons" in Mahalanobis', 1964, tribute). Figure 4.3 shows the list of attendees at that meeting, with some eminent names such as Fisher, Haldane, Frank Yates, Egon Sharpe Pearson, to name a few. By-laws stipulated there would be a Vice-President (regional president), regional secretary, and regional treasurer elected to one-year terms but eligible for re-election, plus six ordinary members on staggered three year terms who are not eligible for re-election until at least one year after completing their term. These officers along with members on Council constituted the Regional Committee.

In 1965, the Region wanted to increase the dues for regional members to match those of North American members but with the proviso that the

FIGURE 4.2
John William Trevan (1887–1956).

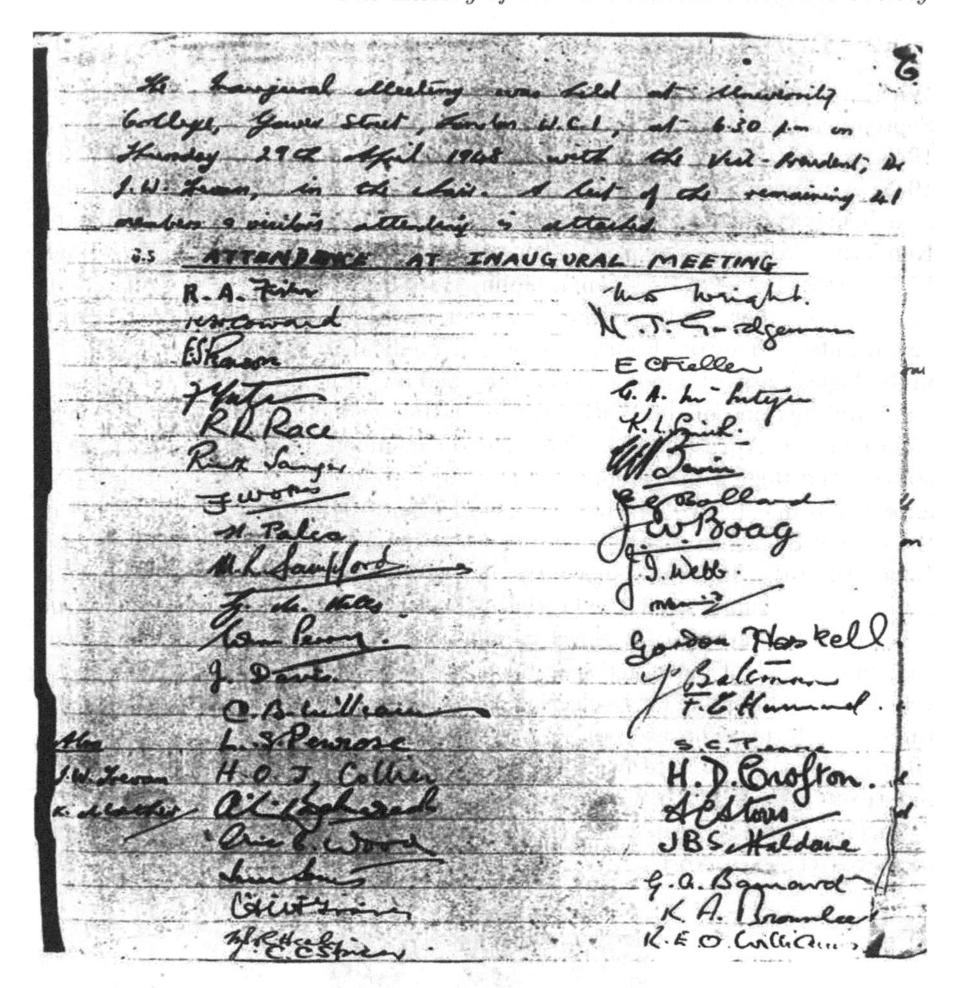

ATTENDANCE AT INAUGURAL MEETING

R. A. Fisher

R. R. Race

L. S. Penrose

H. O. J. Collier

J. W. Boag

Gordon Haskell

H. D. Crofton

JBS Haldane

G. A. Barnard

K. A. Brownlee

FIGURE 4.3
Signatures of attendees at inaugural meeting April 29, 1948.

additional dues remain in the regional coffers. To the Society leadership, this had to be a welcomed relief in sharp contrast to the continual pleas from the Indian members that dues be reduced (see Chapter 5, Section 5.1).

Through 1967–1968 and into 1969, the Society was occupied with unrest within the Region, when first under then Regional President John Gordon Skellam (who in October 1968 wrote "In view of the rebellious mood of the British Region at this moment ...") and then under the next president S Clifford Pearce, the Region became quite concerned about an apparent lack of communication between the Council and the Region. This included a desire for greater cooperation between regions as well as complaints about perceived relations between the Society and ISI. [It is noted that the ISI met in London in 1969, under British organizational leadership.] Lots of long letters were exchanged, with Skellam taking "serious objections" to how things were being handled. Cox was aware of problems in communications and cautioned for patience as she had a committee working on recommendations regarding these issues to be applicable for all regions; this included discussions about setting up a new business office arrangement intended to alleviate some of these problems. Finney came to the rescue providing Skellan with some background, to the point that (in December 1968) Skellan effectively apologised, expressed an appreciation of Cox's wisdom and that he would let things run their course. As an aside, Skellan had beautiful handwriting; see his most gracious letter to Cox in Figure 4.4. This issue led to better advice and procedures to newly appointed officers and Council members.

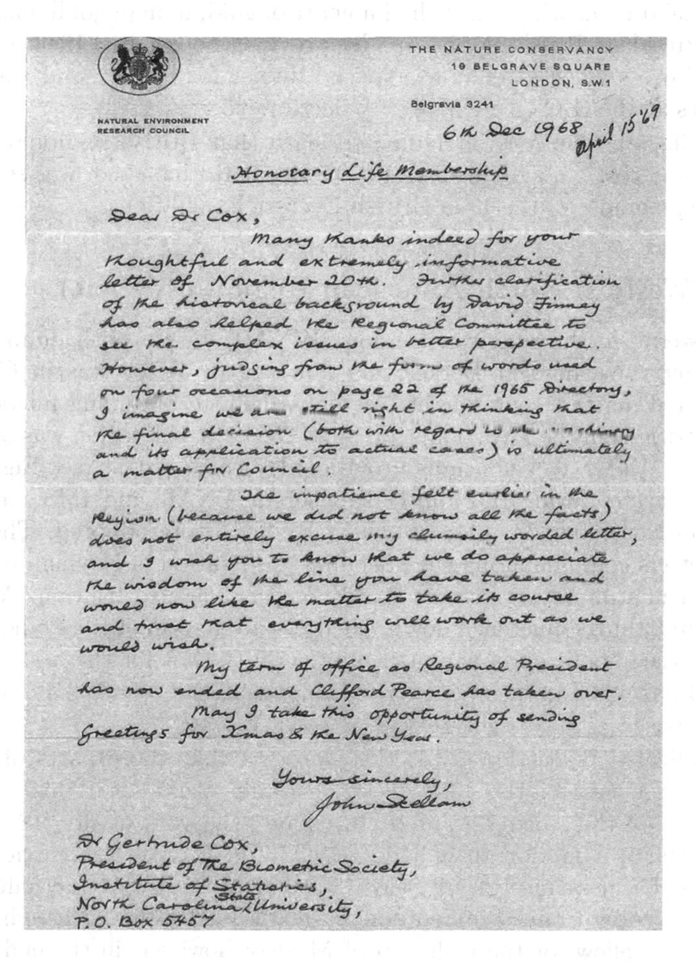

NATURAL ENVIRONMENT
RESEARCH COUNCIL

THE NATURE CONSERVANCY
19 BELGRAVE SQUARE
LONDON, S.W.1

Belgravia 3241

6th Dec 1968
April 15 '69

Honorary Life Membership

Dear Dr Cox,

Many thanks indeed for your thoughtful and extremely informative letter of November 20th. Further clarification of the historical background by David Finney has also helped the Regional Committee to see the complex issues in better perspective. However, judging from the form of words used on four occasions on page 22 of the 1965 Directory, I imagine we are still right in thinking that the final decision (both with regard to the machinery and its application to actual cases) is ultimately a matter for Council.

The impatience felt earlier in the Region (because we did not know all the facts) does not entirely excuse my clumsily worded letter, and I wish you to know that we do appreciate the wisdom of the line you have taken and would now like the matter to take its course and trust that everything will work out as we would wish.

My term of office as Regional President has now ended and Clifford Pearce has taken over.

May I take this opportunity of sending Greetings for Xmas & the New Year.

Yours sincerely,
John Skellam

Dr Gertrude Cox,
President of The Biometric Society,
Institute of Statistics,
North Carolina State University,
P.O. Box 5457

FIGURE 4.4
John Skellam's hand-writing – gracious letter to Gertrude Cox, December 6, 1968.

Skellam himself was a real worker for the Society. Apart from the aforementioned issues that exercised the Region's attention during his presidency, he along with David R Cox wanted the Society to become engaged with a contemporary effort on statistical ecology. The Society subsequently appointed a Liasion Committee on Statistical Ecology with Skellam as Chair to study these efforts.

In a nice lesson in decorum when then Regional President John Gaddum was knighted in the British 1964 New Year Honours List, Finney wrote to Secretary Le Roy with the news and advice that one now addresses him as "Dear Sir John" emphatically **never** as "Dear Sir Gaddum" (or versions thereof). This gentle touch was emblematic of the respect paid to various officers when such honors came their way; the Society may have been a professional organization, but it had a heart of gold, human gold! This evoked Society President Bliss' enquiry in the previous September 1963 as to how he should dress when attending a special dinner for Society and Cambridge dignitaries at the IBC (see Chapter 9, Section 9.5).

On a lighter side, it is intriguing to learn that British Regional member Mike Westmacott, an agricultural statistician at Rothamsted, was selected as a climbing member of the 1953 British Everest Expedition.

4.2.3 Western North American Region (WNAR)

The Western North American Region (WNAR) began with John Tukey's visit to Jerzy Neyman at Berkeley in June 1948. Tukey was in California on unrelated matters, but sought Neyman's help in obtaining introductions to biometricians in the area. Thus, on June 25, 1948, the region was organized with Frank Walter Weymouth as provisional Vice-President (see Figure 4.5). They met again during the jointly sponsored WNAR and IMS meeting in Seattle on November 27, 1948 at which By-laws were adopted. There were several letters complimenting the generous cooperation of Neyman and Allan Birnbaum in stimulating interest in seeing this come to fruition. After this initial spurt, things quietened down much to the chagrin of Bliss who by 1950 was concerned about an apparent lack of activity (not for the first time, nor last, as Bliss' exuberant energy was often not matched by officers in this or in other regions).

As indicated in Section 4.2.1, the plan was that those parts of Canada and the USA west of the 104° West Latitude would form this region. In anticipation of their being granted statehood (the vote of the US Congress came in 1959), a motion to include Alaska and Hawaii was carried at the June 1956 business meeting; it was 1959 before they were actually added (and approved by Council in September 1960). The By-laws were changed in 1960/1964 to allow for the inclusion of Mexico; however, in the end, Mexico decided to form its own national group (see Chapter 6, Section 6.8).

Terms of office started out originally in mid-year until 1952 when Bliss asked the region to change the tenure of officers and the start-date to

FIGURE 4.5
Frank Walter Weymouth (1884–1963).

January to be consistent with other regions; this was readily approved by the membership. Revised By-laws of 1964 established the office of regional president-elect who would serve as regional president the following year and thence a year as regional past-president, a Regional Committee of six ordinary members and regional officers and with international officers including Council members as ex-officio members. The regional treasurer and regional secretary would be elected for one-year terms but could be re-elected. This replaced the 1960 By-laws which had specified two-year terms for the regional president (who could not serve two consecutive terms), and three-year terms for regional secretary and regional treasurer, while the six regional committee members continued in staggered three-year terms. Annually, a nominating committee was appointed by the regional president to prepare a slate of nominees. Usually because of varying circumstances, the regional secretary and regional treasurer offices fluctuated between these being one person for both positions and two persons for the separate positions; the 1960 amended By-laws allowed for both options.

With a financial surplus in 1956, discussions began soliciting ideas on their use; definitive recommendations were to be brought to the 1957 annual meeting. Student awards consisting of a one-year student membership in the Society began in 1961.

As for ENAR earlier in the year, Council approved changes to the By-laws in 1965 so that the regional secretary and regional treasurer officers could be combined since ASA would now be handling much of WNAR's business. This move necessitated an increase in dues to cover the associated cost and also required Council approval.

An intriguing snippet (reference to which came up more than once) referred to a Western Biometric Society begun by Jerzy Neyman in the 1930s. It was short lived because of the intervention of World War II when scientific activities were suspended.

4.2.4 Australasian Region (AR)

The Statistical Society of New South Wales (SSNSW) was formed in November 1947. In the months prior to this society being organized, Helen Newton Turner communicated with Bliss regularly, identifying potential members for "his" Society. By early 1948, Edmund Alfred (Alf) Cornish (see Figure 4.6) began organizing an Australasian Region (AR), covering both Australia and New Zealand, with Turner as acting regional Secretary-Treasurer and Cornish as provisional Vice-President. As for ASA's Biometric Section and ENAR, members of SSNSW would also be early members of AR, so fine lines of operational distinction required special attention, though this fact actually enhanced both entities.

A feature of this region was that each state in Australia and New Zealand, called 'Branches', would have a 'local' convenor starting with Turner, Cornish, and Rupert T Leslie as the New South Wales, South Australia, and Victoria representative, respectively. Bliss' (August 1948) response was that maybe there could be autonomous subregions developed; this approach was however not adopted, though affiliation with national organizations was adopted.

FIGURE 4.6
Edmund Alfred (Alf) Cornish (1909–1973).

Regional secretary Turner reports to Bliss in September 1948 that branch (state) meetings have been held. The Region was formally established in December 1948 with thirty-seven members, with the first Regional meeting held in Melbourne on January 8, 1949, at which the Regional By-laws were discussed by the thirty plus members in attendance. At this time, it was decided to have a regional meeting every two years, supplemented by the various state meetings.

During these early years, frustrations surfaced with slow mail services between Turner (based in Australia) and Bliss (based in the US, though Bliss also acknowledged slow responses for unspecified reasons from other regions especially ENAR and later WNAR), most especially since mails did not reach Turner in sufficient time to allow for the Region to respond in a timely manner. This problem of the vagaries and reliability of surface mail services first appeared in this context; subsequently, later mailings between countries were sent by air.

In a different direction, Bliss was a man of action, one who was very energetic in promoting the growth of the Society. Thus, it was that at first he became concerned about progress when, in addition to the slow mails, Turner also had to deal with her own illnesses and with caring for aging parents. Naturally, regional matters were not front-burner concerns much to Bliss' chagrin. Again, this was a 'first' for the Society; however, over time, when other officers, one of whom was Bliss himself in 1950, be they regional or Society, had to deal with "life goes on" matters demanding to be handled, compassion and sympathy for those involved pervaded the relevant Archive pages, including when someone "injured my left foot" [!].

Cutting across all these frustrations were frequent letters during 1950 between Maurice Belz and Bliss (they had become acquainted when Belz was on leave in the US in 1947, and recall from Chapter 2 that Belz had chaired the Saturday morning session at Woods Hole) with Bliss seeking help and Belz generally quite critical of the lack of involvement, from his perspective, of both Cornish and Turner. Interestingly, however, when the time came, Belz recommended that Cornish be re-elected to a second term but that Turner be replaced; Turner was happy to step down given her extensive family duties, it seemed.

A regional vice-president (who had not earlier served as regional president) officer was added in the 1985 revisions to the regional By-laws to serve in a first and fourth year bracketing a two-year term as regional president, somewhat akin to the terms of the international office. The region further provided for a vacancy in its regional presidency by filling it with its regional vice-president.

5

Growth with More Regions 1949–1958

General principles applicable to all regions were outlined in Chapter 4, Section 4.1. Chapter 4, Section 4.2, described the formation of regions within a year of the beginnings of the Society. In this chapter, regions that emerged during the next ten years are described.

5.1 Indian Region (IR)/Group (GInd)

The Indian Region (IR) was organized during the Indian Science Congress Association meeting in Allahabad with unanimous approval on January 5, 1949. The renowned Prasanta Chandra Mahalanobis (of multivariate Mahalanobis distance fame) was elected as its first Vice-President; see Figure 5.1. With forty plus members, the regional Constitution and Rules (By-laws) were submitted in March 1949 for Council approval. In an interesting twist, at the time, Mahalanobis asked Chester Bliss if he could "back year" his dues as he wanted "to be counted as one of the foundation members"; there is no record telling if Bliss obliged on this question, though there is a receipt seemingly acknowledging this payment.

FIGURE 5.1
Prasanta Chandra Mahalanobis (1893–1972).

DOI: 10.1201/9781003285366-5

The first set of By-laws (1949) specified a Vice-President (Regional President), regional secretary, and regional treasurer for one-year terms but eligible for re-election. These officers were part of the Regional Committee along with nine elected ordinary members and any other member on Council.

Soon, however, as early as 1950, the Society was concerned about non-payment of dues and the expressed wishes of the IR that the dues be decreased, and that without relief, the existence of the region was threatened; more below. The region "collapsed" on February 16, 1951. This caused considerable consternation. Even as this move was being considered, Bliss and also Ronald Fisher tried valiantly to keep the region viable, most especially in light of a pending Symposium at the Indian Statistical Institute in Calcutta. Their pleas went unheeded as the region disbanded. Throughout, Bliss was relentless in trying to see the region function well; at one stage, he even proposed a (particular) candidate for Council because Bliss thought "he might be able to break through the lethargy in India which so far we've been unable to crack."

Later, in July 1952, Bliss wrote to Vinayak Govind Panse, outlining the earlier problems leading to the disbandment of the region, but with now only seven members could Panse encourage three more folk to become members so as to have enough to be reactivated as a national group. There were some concerns about back payments and re-instatements, but Panse was successful, so in January 1953 the Indian National Group (GInd) was formed with Panse as National Secretary.

For four years, 1960–1963, A R Roy served as National Secretary. While there is no evidence that Bliss is not happy with Roy per se, there is apparent frustration that the Indian National Group had not converted back to being a region. Several letters were sent to various Indian members (such as Panse, Mukerji, and Mahalanobis) seeking their help in re-attaining this regional status. Indeed, Bliss, who was ever alert to opportunities to create growth and regions, invoked the role that Fisher had played in the early days encouraging the National Group to return to regional status in Fisher's honor at the time of Fisher's death. In particular, Fisher had responded to Mahalanobis' invitation that the Society be represented at the December 1956 twenty-fifth anniversary of the Indian Statistical Institute and Fisher had stayed on for a few weeks to give lectures. It is December 1963, and Bliss is ending his presidential term, but he continued relentlessly working for members and regions nevertheless.

Cutting across these efforts to re-establish the region, we learn that in 1963, the Indian government set up an Indian Committee for Statistics whose object was to promote international cooperation in statistics and to liaise with international statistical associations (including the Society and ISI), with K R Nair as its secretary. This appeared to be an attempt to pre-empt the regional re-formation. After apparently little progress, this Committee was reconstituted in 1966, with the same members (C R Rao, P C Mahalanobis, S N Bose, V S Huzurbazar, S S Shrikhande, D B Lahiri, V G Panse, and K R Nair).

However, despite all these concerted efforts, it was not until 1989 that the region was re-established (with sixty members), after efforts by then-President Jonas Ellenberg and Secretary Roger Mead along with the Indian National Group Secretary Girja Kant Shukla who had served as National Secretary from 1977 onwards; Shukla was elected as the new regional secretary. Pandurang Vasudeo Sukhatme (known for his work in agricultural sampling) was the new Regional President; it is interesting to note that Sukhatme was a member of the region in its earlier formation back in 1949 and even served on its first Regional Committee.

The issue of dues became a continual theme across the 1950s and 1960s. Indeed, the issue had started with Mahalanobis' letter to Bliss in December 1949 seeking concessions for the Indian members suggesting that allowances be made for differences in real incomes. A backdrop here is the 1949 devaluation of the Indian rupee. Bliss (November 1950) reminds Mahalanobis that the region had asked that its subscriptions to *Biometrics* be suspended and had paid no dues in 1950. Bliss effectively denied the request for dues reduction pointing out that the Society does not allow for reductions of one region over those elsewhere as a question of equity, but that substantial increases were pending for American members and hoped the Indian members would understand in sharing the hardships of non-American members. [Indeed, a sixty percent increase in dues for Western Hemisphere members was approved, while dues for Eastern Hemisphere members remained unchanged in order to equalize the cost all round.]

In that same November 1950 letter, Bliss spent considerable time discussing the upcoming International Indian Symposium (with Mahalanobis as organizer) planned for December 1951 which presupposed a functioning region, and so he appealed to the Indian members to re-dress the situation. The Indian members were adamant however and did not communicate with the Society at all. By January 1951, there appeared to be some confusion in that apparently some members had paid their dues (to the Indian regional secretary) but that these (at least the international portion, that covered primarily the journal costs) had not been forwarded to the Society Secretary-Treasurer. Bliss' secretarial assistant (I N Fisher) wrote that "... as we have had no word from either the Secretary [Indian regional secretary Rao] or from Dr. Mahalanobis we cannot draw a definite conclusion yet." Even as late as November 1951, when discussing costs of the Symposium *Proceedings* with then-President Arthur Linder, Bliss suggested that if Mahalanobis has any remaining funds that he use them "in re-activating the Indian Region"; Bliss never gives up! At the time of the Symposium, discussions were held between Fisher, Frank Yates and Linder with Panse and C R Rao on the revival of the IR hopefully with Panse as regional secretary, but that the decision had to come from the Indian members themselves.

It is unclear when the region formally disbanded though the operative date was February 1951 when a report informed Society officers that the Indian regional secretary Rao had requested members who wished to continue in the Society transfer to being At-Large members (where the dues were lower) "pending further negotiations"; Rao's February 1951 letter to Secretary Bliss implied this was back in December 1949. The Society thus proceeded with back billings for 1950 and 1951. Ever hopeful, Bliss wrote to Linder in February 1951 that with the "continued inactivity of [the region] .... Unless we have a favorable reply from Mahalanobis or Rao, [he, Bliss] proposes the Indian Region be placed on the inactive status in view of apparent lack of interest ... ." Bliss also hoped that with Linder's plans to work in India July to December 1951 and with the planned Symposium in India in December 1951, that somehow the region could still be reactivated. These efforts were unsuccessful.

In the 1960s, this issue continued unabated. The "problems of the Indian members ... were considered at some length" during the Cambridge IBC in September 1963. Council approved the appointment of A R Kamat as the next National Secretary for the Indian National Group (after four good years of service from A R Roy because Bliss thinks turn-over is healthy), starting in January 1964; Council hoped Kamat would also work toward the re-formation of the region. Apparently, many dues had lapsed over 1963–1964, and so one of Kamat's first tasks was to collect these unpaid dues. The discussion, over what dues should be, never ceased, however. Kamat was at least persistent and consistent in his efforts on behalf of the Indian members. At first, there was some sympathy, even to the extent that Bliss when approached directly by Kamat thought a reduction to US$4 might be possible. Apparently, Kamat took Bliss' comments as fact and so the dues were assessed at $4 for 1966 and 1967 but were to return in 1968 to the properly assessed value (which was then $6 assuming no further increases). The requests and discussions however continued. The Finance Committee thought otherwise as they tired of the continual requests. Kamat appealed to Assistant Treasurer Robert O Kuehl in 1966 seeking special concessions in light of the Indian rupee devaluation. By 1967, Treasurer Henry Tucker, pointing out the persistent history from India and the fact that Brazil has managed despite their devaluations, was also not inclined to make concessions. He was especially concerned that concessions to one region (here India) could invite similar requests from other regions, and that, in particular, dues were needed to protect the publication of the journal *Biometrics*. Indeed, the *Biometrics* editor (Michael Sampford) opined that they should not make a special case for Indian members and wondered if these members would be more suited to being associate members. The March 1969 Executive Committee meeting emphatically turned down Kamat's request that the Indian dues remain as US$4 since by now the properly assessed US$6 was the same as student dues in ENAR-WNAR, other countries were monetarily stressed too, and there was a need to be "fair" to all members across regions/groups. On completion of Kamat's four years term, he was replaced by G R Seth on January 1, 1968, and then M S Chakraborty was acting secretary in late 1969 while Seth was on leave and then replaced Seth in 1970; but the requests for dues reductions continued unabated.

5.2 Région Française (French Region, RF)

The Région Française (RF, French Region) was formed in March 15 1949 in Paris. To satisfy a 1901 French law governing French organizations, members simultaneously formed the Société Française de Biométrie and required members of this latter society to be members of the French Region also. The first Vice-President was Maurice René Fréchet (Figure 5.2a), a distinguished mathematician best known in the statistical world for the Fréchet distance measuring the similarity between curves.

With the formation of the French Region, the flurry of activity between various parts of Western Europe to form a region to include France, Italy, Switzerland, and eventually Benelux, came to an end. That concept essentially took root with Italy's Adriano Buzzati-Traverso's letter to Secretary Bliss in February 1948 telling him that he had been in contact with Georges Teissier (of France) with definite proposals to establish a single region. There are some fascinating, but brutally frank, assessments of possibilities addressed in the exchanges of letters. Emotions were clearly still raw, and so political realities were being cautiously, and not so cautiously, adjudged. Remember World War II had only ended in the European theater of operations in mid-1945, and this is early 1948. Benelux felt the existing "difficulties for the reunion of all the western European countries into one single region" were essentially too much; so that by December 1948, proposals for a joint French-Italian region emerged with separate secretaries (morphed into a joint secretary-treasurer by February 1949) for each country; the first such proposed Daniel Schwartz

FIGURE 5.2
(a) Maurice René Fréchet (1878–1973) and (b) Egon Ullrich (1902–1957).

for France and Luigi Luca Cavalli-Sforza for Italy. Buzzati-Traverso, despite his prominent role in fostering the joint venture, deferred, since he was too busy with organizing a genetics conference. Buzzati-Traverso hoped that, with the upcoming Second IBC to be held in Geneva in 1949 (see Chapter 9, Section 9.2), Swiss members would also be inspired to join this single region. Ultimately, France and Italy became their own separate regions; see Section 5.4 for the growth of the Italian region.

The first set of Regional By-laws called for a Vice-President (Regional President), regional secretary and regional treasurer elected to three-year terms; the Regional President could not be re-elected until at least three years later, but the regional secretary and regional treasurer were eligible for immediate re-election. The Regional Committee consisted of three ordinary members who served staggered three-year terms, as well as Council members. The region introduced its first Young Biometricians Meeting in 1997.

5.3 Belgian and Belgian Congo Region; Belgian Region (RBe)

Early in 1952, Léopold Martin proposed plans to build up a Belgium section to be pooled with members from The Netherlands and Luxembourg to form a Benelux region. In the end, in October 1952, Council unanimously approved Bliss' recommendation to appoint Martin as the National Secretary of a Belgium and Belgium dependencies (Belgium Congo) Group. Martin had been particularly effective in gathering forty-three new members for the Society. However, in December 1952, the Société Adolphe Quetelet was formed as the Belgian Region (à la the French Region), in that to be part of the Société Adolphe Quetelet, an individual had first to be a member of the Society.

Interestingly, the English translation of their statutes included a comparison of the French and non-lucrative Belgium law. Paul Spehl (Figure 5.3), a medical biostatistician, was the first Vice-President. Martin continued as regional secretary for ten years and continued to be an effective member of the Society becoming President for 1960–1961.

Over the years, the region has been active with many regional meetings, as well as contributing much to the international Society and its committees and general officers. This included organizing the 1988 IBC in Namur (see Chapter 9, Section 9.1.14). One distinguishing feature of the region was its publication *Biométrie-Praximétrie*, founded by Martin. This was a Society journal approved by Council in June 1959 with the first issue in 1960. Publication ceased in 1994 (see Chapter 8, Section 8.4).

The Belgian and Belgian Congo Region changed to the Belgian Region (Société Adolphe Quetelet) in 1960 when the Belgian Congo became an independent country.

FIGURE 5.3
Léopold Martin (1909–1991) listening to Paul Spehl (1887–1980).

5.4 Italian Region (RItl)

As indicated in Section 5.2, in February 1948, the Italian Adriano Bussati-Traverso had conceived of forming a region involving France, Italy, Switzerland, and Benelux, and so initiated contact with the Frenchman Georges Teissier about this. Later in November 1948, Bussati-Traverso was still optimistic suggesting to Bliss that, despite the Dutchman Neurdenburg's objections (see Chapter 6, Section 6.1), things were improving in Europe and that such a region could still be conceivable and workable. Nevertheless, after meeting with Teissier, still in 1948, he instead proposed a French-Italian region with presidents alternating between the two countries yearly and with a separate secretary-treasurer for each. However, with the formation in 1949 of the French Region, these proposals became moot. Thus in 1949, Italian members became a National Group (GItl) with Bussati-Traverso as its first National Secretary.

Nevertheless, efforts to form a separate Italian region continued. Letters were exchanged between Bliss and Bussati-Traverso from late 1949 to early 1951 encouraging formation especially as there were now over forty members of the Society based in Italy (fast approaching the fifty needed at the time to form a region). In February 1951, Luigi Luca Cavalli-Sforza took over as the acting secretary of the still-to-be-official region. Cavalli-Sforza was now the key person for the region's establishment – and he was a busy man, being the organizer for a meeting in Milan and later the organizing secretary for the Third IBC slated for Bellagio in September 1953 (see Chapter 9, Section 9.3).

FIGURE 5.4
Claudio Barigozzi (1909–1996).

Progress toward a formal establishment of the region was slow; Bliss was persistent, however, even to the point of asking in May 1952 if there was any likelihood the region would be established before the 1953 IBC – Cavalli-Sforza hoped so!

Finally, in 1953, the region became a reality, and Cavalli-Sforza was able to report to Bliss in June 1953 that elections had occurred with Claudio Barigozzi (Bari), prominent in Italian genetics activity, as the inaugural Regional President (see Figure 5.4). Cavalli-Sforza stayed on as regional secretary, a role he continued until the end of 1957.

At its formation, the Regional Statutes (By-laws) provided for a Regional President (at the time also a Society Vice-President), regional secretary and regional treasurer and three ordinary members who together comprised the Regional Committee. Officers were elected to two-year terms and could be re-elected except that the regional president could not be re-elected for two consecutive periods. Members of Council from the region were also on the Regional Committee.

Since the region was founded mainly by geneticists (primarily from agriculture; medical biometricians/geneticists joined later), their influence was considerable. They were particularly active in organizing and sponsoring symposia on related topics attracting international participation. This included sponsoring the third IBC in Bellagio in 1953, and the Varenna Seminar in Biometry in 1955; see Chapter 9, Section 9.1.3 and Section 9.2.3, respectively.

5.5 Deutsche Region (German Region, DR)

In June 1952, Bliss suggested to Maria Pia Geppert that she be appointed as National Secretary for the four German members of the Society,

(a)

(b)

FIGURE 5.5
(a) Maria Pia Geppert (1907–1997) and (b) Joachim-Hermann Scharf (1921–2014).

subsequently approved by Council in September 1952. Concurrently, Bliss was communicating with Richard Prigge who was being successful in enrolling more members. By January 1953, there were enough members to form a National Group; by May 1954, there were enough (forty-three) to form a region. Geppert worked on preparing Regional By-laws, with the Statutes approved in January 1955. Though Prigge was elected as the first Regional President, he declined citing illness. New voting selected Egon Ullrich, a mathematician, who took up office as the actual first Regional President in March 1955 (see Figure 5.2b); he died suddenly in office in mid-1957. While she herself was never elected Regional President (although when still a National Group during 1952–1954, she was regarded as a de facto president), Geppert was an ever-present figure capably serving as regional secretary and/or treasurer on many occasions (see Figure 5.5a). In addition, Geppert along with Ottokar Heinisch co-founded the journal *Biometrische Zeitscrift* in 1959 as a regional publication (see Chapter 8, Section 8.3.2); and she was the first German member to become an Honorary member of the Society. In the region's fortieth anniversary history, Geppart modestly said that their halo image of her should be destroyed [!].

The 1955 By-laws held that officers be elected for one-year terms, and that no Regional President could serve more than two consecutive years but that there were no time limit restrictions for other officers. The regional secretary and regional treasurer positions could be combined. The Regional Committee consisted of three to five members in addition to regional officers. The Region hosted the 1970 IBC in Hannover (see Chapter 9, Section 9.1.7) under the organizing guidance of Berthold Schneider (President, 1970–1971).

With the construction of the Berlin Wall, the question of how to incorporate East German (German Democratic Republic) members was a somewhat tricky dilemma. In August 1962, in what appeared to be an attempt to mitigate difficulties of currency restrictions between East and West Germany, Heinisch (then the immediate past German Regional President and an editor of *Biometrische Zeitschrift*) wanted to allow East German members the option to swap *Biometrics* for *Biometrische Zeitschrift*. Bliss was disinclined to effect any such swap partly because he thought it would be impracticable, and wrote to Sampford (*Biometrics* Editor) that *Biometrics* addresses should not be shared with *Biometrische Zeitscrhift* staff. There was a series of letters between Treasurer Marvin Kastenbaum, Heinisch, Sampford and Bliss over the next four months. They were clearly sympathetic to the currency burdens exacerbated by the fact that the German Region's treasurer Geppert lived in West Germany and Heinisch lived in East Germany.

The concerns culminated in a long detailed December 1962 letter to Heinisch, in which Bliss carefully set out possibilities and non-possibilities. In particular, since *Biometrische Zeitschrift* was owned by a commercial publisher (Akademie-Verlag, based in East Berlin) and not by the Society, Society members could not facilitate any subscription collections for that journal. However, to ameliorate the financial constraints of East German members, Bliss suggested that some members could become Associate members (who would not receive *Biometrics*) with such members strategically placed geographically so that they have access to *Biometrics* from members to share the journal (but he thought that 20%–25% of the East German members should nevertheless stay as full Society members); in this case, the Associate members would be assessed the international portion of the dues ($0.75), whereas the full members would pay the international portion plus the *Biometrics* portion ($2.75). This was seen as a temporary measure with a return to full status for all members when the "present restrictions on currency exchange are relaxed." Later efforts (in 1964) to consider photographic reproduction of the journal, not just for East Germany but for all East European members, were abandoned since it would require an impossibly high totally unattainable number of members to make it cost effective.

A new direction became necessary when the elected Regional President Herbert Jordan, a resident of East Germany, was forbidden to take office by the East German government. The seat thus officially remained vacant for the year (1966), although Jordan was a type of substitute president actually active and with the assistance of the preceding Regional President Siegfried Koller.

Based on an initiative of Erna Weber from May 1968 to form a separate East German group (Deutsche Democratic Republic (DDR)), communications between Secretary Henri Le Roy and President Gertrude Cox continued regularly in the later months (September-October-November) wondering in particular about how such a formation fitted with (West) German Regional By-laws, and about how East German scientists living in other countries when Society regions were supposed to be "delimited geographically." The Society leadership was anxious to hear from the (West) German Region their views on there being a separate East German Region. Le Roy is concerned that the region would be "more political than scientific" and suggested that safeguards against this be written into the By-laws, including a requirement that only scientists can be members. The By-laws were proposed in October 1968. By November 1968, Le Roy assured Cox that, despite their concerns about underlying political questions, he thought the two regions had a right to, and could, co-exist without problems. Eventually, the new group was formally approved by Council in May 1969, with Erna Weber as the inaugural National Secretary. A review (in October 1968) of proposed By-laws for a new East German group suggested that the present "German region" members would stay as such until their current membership expired at which time they would be transferred over to the new East German group.

That was in 1969. By 1971, the group wanted to be a region and so quickly prepared Regional By-laws, for this Region of the German Democratic Republic (DDR-Region, or Region of the GDR) and sent to Council for approval. Professor Joachim-Hermann Scharf (Figure 5.5b) became its first Regional President. Since the membership fees were paid by public agencies, there was a limitation on the official size of the region (set at fifty), with new members only being added when a current member retired; this prevailed until 1991 when the two regions merged.

Part of the German Region story is the merger in 1991 of the West German Region with the German Democratic Republic (East German) Region (RGDR) after the fall of the Berlin Wall in 1989. The Council vote on dissolution of RGDR came quickly, in December 1990. Thus, it was that the year 1991 saw the reunification of the two regions.

The Region's fortieth anniversary featured a special three-day Colloquium held in March 1993; see Figure 5.6. This was more than an important anniversary, however; it was also a joyous celebration of the union of the two regions. Thus, scientific talks were held at the Free University (in the former Western Germany), with an Opening Ceremony at a Symphony Hall

FIGURE 5.6
Program cover – fortieth anniversary and merger of both German Regions.

next door to Berlin's Humboldt University (in the former Eastern Germany). This Opening Ceremony began with a live performance of Beethoven and closed with a live performance of Mozart! In between, there were numerous congratulatory messages including one from Lynne Billard (then Society Vice-President representing Society President Niels Keiding). After Mozart, participants strode triumphantly to a restaurant where RGDR members, forbidden to hold scientific meetings, would over the years gather clandestinely to discuss their science over dinner. The author (Billard) was startled to discover this restaurant abutted the main train station; this evoked stomach-churning memories of being kept (no disembarking allowed) on a train in 1969 while armed guards and dogs thumped across the top of the train cars and walked along the outsides searching for people hoping to escape. The concept that there could be a return to this setting in happier circumstances was too foreign to contemplate. More details of this celebration can be found in Billard (1994).

As a personal aside, while at the time the author in June 1969 was preoccupied with the jarring sight of armed guards and sniffing guard dogs at the Berlin train station, it is poignant to realize now that outside the strictures of that train station, science was at work in the concurrent formation of a National Group of the Society. Juxtaposing these extremes is too mind-boggling to grasp fully.

5.6 Região Brasileira (Brazilian Region, RBras)

As early as January 1954, there was interest among members in Brazil and news that a National Secretary would be appointed "soon," this followed Bliss' approaches to Américo Groszmann in late 1953 as to his interest in such an appointment. Groszmann agreed as long as Brazilian members dues were set at the same level as for European members.

Meantime, ever opportunistic, Bliss focused on the fact that the ISI was to hold its twenty-ninth Session in Rio de Janeiro Brazil in 1955. In what was by now a regular pattern, Bliss therefore wanted to hold an accompanying conference in order to promote membership and the formation of a region in Brazil. After a long series of letters to and fro, eventually the Society settled on a Biometric Symposium held in Campinas over July 4–10, 1955; see Chapter 9, Section 9.2.2).

The Symposium was very successful with many international and national participants in attendance and distinguished itself in many ways. More importantly however, this Symposium generated a lot of local interest and, as hoped, really hastened the formation of the region; members enthusiastically endorsed this step at a business meeting during the Symposium, and regional By-laws were established in September 1955.

Thus, the Brazilian region (RBras) officially came into being in January 1956 when Council approved its regional By-laws (though accommodation to national legal requirements for incorporation of a region would necessitate some further amendments). One requirement of the Brazilian law was that no foreign word or phrase could be used in its name; thus, members opted for "Região Brasileira da Sociedade Internacional de Biometria" adding the word "Internacional" to the English translation following the precedent set by the French Region. The Symposium's local organizer Constantino Gonçalves Fraga Jr (an experimental design and agricultural statistician) became the first Regional President. Groszmann was elected as the inaugural regional treasurer, and was duly complimented in increasing the membership from ten to fifty-three members in the intervening two short years.

Officers consisted of the Regional President (who was permitted to serve two consecutive one-year terms, if so elected), and regional secretary and regional treasurer which could be combined into one secretary-treasurer position. The governing body was the Regional Commission consisting of the current officers and six other members with one-third rotating off each year. A special committee appointed by the Regional Commission prepared the slate for the annual elections.

Brazil is a vast country geographically. However, most members lived in the state of São Paulo; so it was not surprising that thirty-six of the first thirty-seven meetings in the first thirty-three years were held in that state, with almost half of these being hosted in Campinas. The dominance of this state continued with roughly 80% of meetings occurring within its borders. One exception was when the Region hosted the 2010 IBC at Florianópolis, a coastal city in the state Santa Catarina, having previously hosted the 1979 IBC at Guarujá but in São Paulo state.

As also happened elsewhere, in the late 1950s, huge inflation pressures resulted in considerable financial difficulties in Brazil. The region however reported on these problems in its annual report to Secretary Michael Healy, adding that they hoped to obtain a local grant to remedy the situation. This attitude was a welcome relief to the Society, especially when contrasted with the approach of the Indian members (see Section 5.1). Indeed, the discussions regarding the Indian situation included favorable comparisons with the Brazil members. The Region's webpage proudly asserts (rightly so) that it "has gone through several crises, all of them overcome with relative ease, in such a way that it never stopped its operation" (see Gomes, 1989).

There was a flurry of letters between regional secretary Paulo Mello Freire, Secretary Healy and President Cyril Goulden late 1959 to early 1960 when Freire reported to Healy about some (seemingly completely inadvertent) irregularities in the September 1959 election for a new regional secretary (Freire was not himself a candidate to continue in office). The regional folk were of the opinion that the September results should be vacated with a new election slated for April 1960. In effect, the region was seeking affirmation from Society officers that their proposals were satisfactory. Though the infraction was relatively minor in fact, it is instructive that these Society officers, while agreeing with the regional approach, "were most hesitant to impose any solution upon the Region" such was the reluctance to interfere with regional affairs. Not that overall oversight of the regions was not there. For example, years later (in April 1968), Secretary Le Roy had to remind Frederico Pimental Gomes that his term as Regional President was already past the two-year limit fixed by the constitution, and that he would need to take appropriate steps for a successor; Le Roy was apologetic that he had to write that an "active President of a Region is obliged to retire."

6

Growth from 1959 – More Regions, National Groups and Networks

As the decades rolled around and passed to the next decade, the Society continued to grow. Those regions that were formally formed after 1958 are presented in the chapter. As shall be seen, in many instances, some of these regions had already existed, as national groups, for several years before converting to regional status.

National groups typically consisted of geographical areas with at least ten members, but they were not sufficiently large to attain regional status (which until 2012, required about fifty members). These national groups are included in this chapter. This chapter discusses the early manifestations of networks, usually loosely defined amalgamations of smaller entities coalesced into a single network for the purposes of scientific meetings.

6.1 Afdeling Netherland (The Netherlands Region, ANed)

Andriano Buzzati-Traverso's February 1948 letter to Chester Bliss with his plans, or dreams, of a region to include France, Italy, Switzerland, and the Benelux countries (Belgium, Netherlands, and Luxembourg) sparked a lot of interest, some reactionary, in Western Europe. In the Netherlands, this was taken up by Woods Hole attendee, M. G. Neurdenburg (a medical researcher/ inspector, seemingly an energetic, colorful but ultimately enigmatic personality who strode the stage until 1953).

Starting with his first letter in October 1948, Neurdenburg engaged in a extended sequence of back-and-forth (pages long) letters with Bliss covering a wide range of issues. In response to the immediate question of one Western European region, Neurdenburg pointed out potential language difficulties, problems with dues because of differing currency exchange rates, reference to "Five years of german occupation and three years of recovery" and wondered if a return to international standards was possible, and somehow brought in the need for "settlement of the Indonesian affairs ... and the menace of Asiatic communism ..." He also drew attention to the two statistical organizations that already existed in the Netherlands (one of which had

DOI: 10.1201/9781003285366-6

started the quarterly publication *Statistica*). His discourses continued into 1949, whereby he repeated his concerns about dues collections, problems with there being two societies with whom "negotiations are troubled by personal antipathies" and "unpersonnel members." Bliss advised Neurdenburg that because of the French laws as they impacted France, one Western European region would no longer be possible. Neurdenburg still wrote about these issues regardless, even wondered (in an apparent reference to the 1949 Geneva International Biometric Conference (IBC)) why anyone would go to this IBC since "Switzerland is a very, very expensive country without being a world-wide attraction" though he continued on by saying that if the International Statistical Institute (ISI) Session was also being held this could be of interest. Threaded through Neurdenburg's outspokenness were some real concerns of real points that did need to be considered however.

One big issue revolved around each member having to pay a currency exchange rightly seen as a real imposition. Thus, meantime, Bliss had been proposing the formation of national groups for countries with only ten to twenty-five members as an alternative to regions, with a National Secretary to collect dues and then to send one collective check covering all group members dues payments. Since (it is now July 1949) in a second vote, Council approved this idea but not unanimously (sixteen in favor and four abstentions and/or against) and since Neurdenburg strongly objected, Bliss reported back to Neurdenburg that the idea be shelved until after the Geneva IBC. In addition, despite Neurdenburg's objections, Bliss tried to explain that this should allay some of his/Neurdenburg's criticisms about costs of currency exchange, although at the time three more members would still be needed to form a viable group. Neurdenburg was subsequently appointed as National Secretary with 1949 as the national group's first "organizational" year.

After all those very long letters to Bliss, suddenly there is nothing from Neurdenburg. By November 1950, Bliss was writing asking for the national group's dues and also Neurdenburg's own dues! Without an accounting of who had, or had not, paid dues, no journal issue could be mailed to the members. Neurdenburg replied with convoluted messages and excuses but did indicate some monies would be coming. However, as Bliss replied in December 1950, there was a considerable disparity between their respective interpretations of the Dutch accounts, including a statement that "no funds whatever have been received from [Neurdenburg]." Bliss gave detailed accounting showing what was owed for 1949 and also for 1950, ending (by reference to the relevant Council By-law) with explicit instructions (on various) and a reminder that, without paid-up dues, journals could not be sent to a member and that it would be necessary to "suspend the privileges attendant" upon having a National Secretary. Neurdenburg promptly replied, even suggesting he liked Bliss' ultimatums [really?] but continued with his now-oft-repeated excuses and complaints. There was more back and forth through January 1951. These exchanges were somewhat surreal.

Then, there seemed to be silence until in May 1952 Bliss wrote back to Neurdenburg about not receiving anything from him and offered the suggestion that he (Neurdenburg) step down as National Secretary. Despite all these communications, Neurdenburg continued to complain (e.g., he did not want to spend money on sending out Society ballots), continued to seek Society funds (for various, including stamps, etc.) and continued his outspokenness even "disagree[ing] towards admission of Germans if no special caution is taken." While his was a seemingly outrageous diatribe (not, in the author's opinion, worth repeating here), it should be remembered that this was 1952 when perhaps emotions were still raw and fragile. Bliss in his ever polite manner continued to ask for funds owed and information from the national group; but by November 1952, Neurdenburg had become quite abrasive toward Bliss. Bliss continued to outline the duties of a National Secretary, and in particular sought paid-up dues and membership lists. Bliss also advised him that with the pending formation of a Belgian Region, earlier conversations about a possible Benelux region were now moot.

Bliss knew a change was in order, and so in February 1953 sought advice from T J D Erlee (who was himself confused by events within the group); he cryptically suggested that Neurdenburg, from his infrequent letters, "is finding the responsibilities onerous." Simultaneously, Bliss wrote to Neurdenburg explaining that it was time for "someone else to take over" though Neuredenburg was still asked to clear up the accounts for 1952 and 1953. Erlee was too busy to take on the task himself but suggested E van der Laan who was duly appointed as National Secretary in April 1953 – charged with "clearing up the problems of the Society in Holland and adding new members." Both Bliss and van der Laan were very diplomatic and gentlemanly, van der Lann was most regretful about Neurdenburg's behavior, suggested the controversy be "skip[ped]" and that he move on with the administration. Perhaps not unexpectedly, with van der Laan's appointment, Neurdenburg wrote "indignantly" to Bliss with a long diatribe on (lots of things, disputed by Erlee and van der Laan; and disparaged van der Laan's credentials) and wanted there to be a change in general secretary (Bliss' office), to which Bliss politely told him how he could make any such proposal for consideration at the Council meeting at the upcoming Belligio IBC and/or write to the management with his complaints (about Bliss). Along the way, Neurdenburg named a National Committee for Holland with one of its duties to name his successor; since he was no longer National Secretary, he had no authority to do this (at least not in the name of the Society). However, his committee "did not come off." Clearly, Neurdenburg was not going to go quietly! There were more letters on dues, but eventually in July 1953, Neurdenburg sent in some funds and promised the rest after he returns from his holidays; though still complaining (about things, and vilifying Erlee and van der Laan in particular), with apparent contriteness, he hoped "all [of his] mistakes and misunderstandings are solved .... [and] glad this burden is finished." However, the saga continued since by October 1953, van der Laan reports that Neurdenburg had sent "part of the cash but could do no more yet because he had to go to London." [!]

Van der Laan took over and was a very welcoming breath of fresh air. Where Neurdenburg's letters went for pages, van der Laan was decidedly short

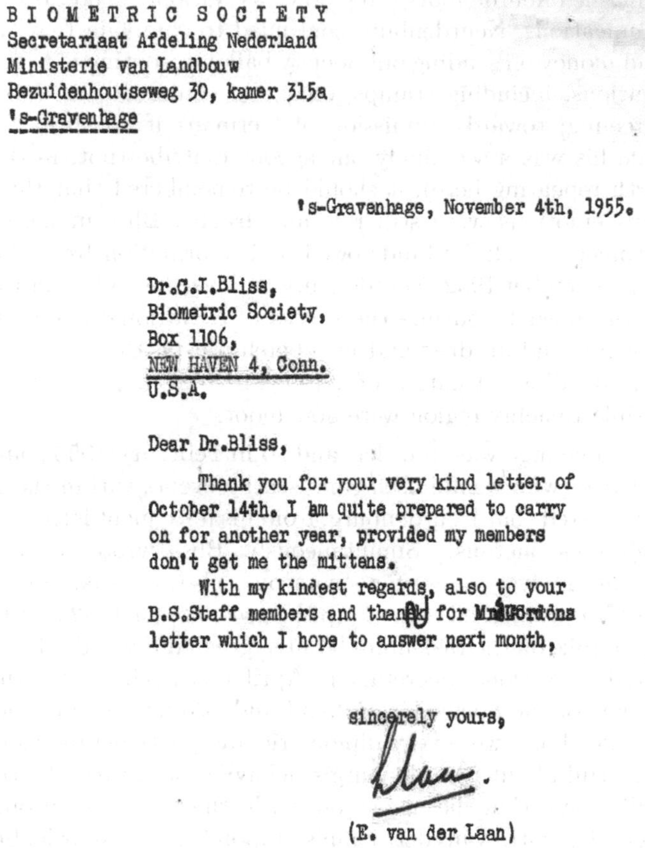

BIOMETRIC SOCIETY
Secretariaat Afdeling Nederland
Ministerie van Landbouw
Bezuidenhoutseweg 30, kamer 315a
's-Gravenhage

's-Gravenhage, November 4th, 1955.

Dr.C.I.Bliss,
Biometric Society,
Box 1106,
NEW HAVEN 4, Conn.
U.S.A.

Dear Dr.Bliss,

Thank you for your very kind letter of October 14th. I am quite prepared to carry on for another year, provided my members don't get me the mittens.

With my kindest regards, also to your B.S.Staff members and thanks for Mr. [illegible] letter which I hope to answer next month,

sincerely yours,

(E. van der Laan)

FIGURE 6.1
E van der Laan wrote short letters.

but to the point; see, e.g., Figure 6.1. He was also colorful but in a different way to his predecessor. For example, paid-up members were "Good boys!", two who "don't answer my letters (are bad boys!)" and so canceled, (the previous earlier twenty-six) paid-up members are "nice people", and so it goes. On his work retirement (at the end of 1954) after Bliss encouraged him to stay on another year, van der Laan was "quite prepared [to do so], provided my members don't get me the mittens." Under van der Laan's tutelage, membership flourished. In 1958, Secretary Michael Healy referred to van der Laan as the "pilot in the vegetable jungle" and urged him to establish a region and to set up relevant By-laws.

Subsequently, the Region Netherlands formally came into being May 17, 1960, where-upon van der Laan wanted to resign as National Secretary. Healy acknowledged this formation as a monument to his long term in office and declared he was indeed the "Father" of the Region; earlier, Bliss had declared

FIGURE 6.2
David Karel de Jongh (1909–1962).

him as "one of our most cooperative and gracious Secretaries" (as indeed it seemed he was). David Karel de Jongh, a medical researcher in pharmacology, was elected as the first Regional President (Figure 6.2).

Through the 1950s nevertheless, the two, then three, biometrical clubs of the Netherlands would meet for scientific presentations. In their December 1955 meeting, due to lack of funds, it was decided to discontinue publication of their small Dutch language periodical *Biometric Contacts* which had run for two years.

For a short time in mid-1963, it looked like the Region had returned to rocky ground with the Treasurer wondering why the Regional President de Jongh had not paid his dues, only to learn that dues had been fully paid up until his sudden death in 1962. No one had informed the Society! The regional secretary was generally unresponsive; "Society affairs in Holland are at a low ebb" which opinion induced Bliss to concur with regional secretary Hendrik de Jonge that determination of de Jongh's successor be delayed until after the Cambridge IBC (slated for September 1963) where it was proposed to discuss how statistical activities should be organized in The Netherlands. There was no Regional President from 1963 to 1976. In 1970, with membership below that required to be a region, Vice-President Gertrude Cox reported on her interactions with the Dutchman Leo C A Corsten who explained that most members were connected with libraries and joined so that these libraries would enjoy a lower subscription rate. There being very few active members, Council felt it was unfair to other regions who had to have fifty members to form regional status if the Netherlands was able to remain as a region with so few members. Eventually, Corsten became regional president for the years

1976–1978; and then the region reverted back to having no elected president until 1991. Throughout these decades, Hendrik de Jonge was the region's rock by stepping in and serving as regional secretary/treasurer from 1960 to 1982.

With a start in 1997, the Region presents a Biometry Award to a refereed publication in a biometrical field published in the previous two years.

6.2 Region Österreich-Schweiz (Austro-Swiss Region, ROeS)

The Austro-Swiss Region (ROeS) had its origins in a Swiss Group when Arthur Linder (second Society President, 1950–1951, as an At-Large member) accepted the post as National Secretary of a newly formed Swiss Group officially in 1954. Austrian participants attending the 1956 Linz Seminar (see Chapter 9, Section 9.2.3) discussed formation of an Austrian section of the Society. Linder was followed by Henri Le Roy (also President 1976–1977; as well as Secretary for the five years 1963–1968, and eventually as a Regional President in 1970–1971). Le Roy was an active member indeed. In particular, he orchestrated the push to unite with Austrian members into the formation of the Austro-Swiss Region. The Regional By-laws and thus the Region were formally adopted in January 1962. Secretary Healy declared that Le Roy was the "principal architect" of the new region.

The first Regional President was Prof Leopold K Schmetterer (a mathematician who became a probabilist and theoretical statistician); see Figure 6.3.

The region continued to think beyond its immediate confines. For example, (in 1971) Le Roy and W J Ziegler, the Society and Regional

FIGURE 6.3
Leopold Karl Schmetterer (1919–2004).

Secretaries, respectively, sought IBS funds to support a lecture series for young investigators. The Region introduced an Arthur Linder Prize for members under thirty-five-years with the first prize being awarded in 1997.

6.3 Japanese Region (JR)

As early as 1952, Bliss reached out to At-Large members in Japan, seeking their interest in forming a national group. Bliss wanted to appoint Matayoshi Hatamura as the National Secretary, but deferred to Hatamura's preference to let the local members decide. The group's first meeting was in August 1953, at which all fifteen members present approved the formation of a region – referred to as the "Chapter of Japan." Dr Motosaburo Masuyama was proposed as the vice-president (i.e., Regional President), but Masuyama declined as he was leaving the country for an extended period; so, Hatamura agreed to stay on as National Secretary, a role he maintained until 1969. When a new National Secretary took up his position in 1970, Hatamura continued to serve but as the national treasurer. From the initial fifteen members in 1952, the membership never stopped growing; there were seventeen reported members in 1963, 81 in 1965, and almost 100 by 1969. In 1969, Secretary Hanspeter Thöni explained to Hatamura that only fifty members are required for a region, to which Hatamura asked for the By-laws of some regions to be used as a model for any Japanese formation. Nevertheless, the membership apparently preferred to remain as a national group. It was not until November 1978 when Secretary James Williams recommended to Council that a region be approved. With that action, the new Japanese Region (JR) came into being at the beginning of 1979 with Chikio Hayashi, renowned for the Hayashi's quantification methods for questionnaire data, as the first Regional President (see Figure 6.4).

FIGURE 6.4
Chikio Hayashi (1918–2002).

6.4 Nordic Region (NR)/Nordic-Baltic Region (NBR)

Although the Nordic Region (NR) came into being in 1982, its formation was preceded by a number of national groups which made up the Nordic geographical region.

The origins began when the Danish immunologist Niels Kaj Jerne (incidentally, a 1984 Nobel Laureate no less) was elected to the 1951–1953 Council. There thence followed several letters between Jerne and Bliss about the desirability, or possibilities, of forming a Scandinavian region. In a March 1951 communication, Bliss offered that having a Scandinavian region was approved at one of the first Council meetings; at the time there was no statutory lower limit to the size of a region. The concept to form a region was approved at a general meeting of Danish members in May 1951. Discussions between Jerne and Bliss over the next twelve months included whether or not the 'dutch' might be interested, whether German members should be included, and even draft regulations for a "Proposed Scandinavian-Dutch Region" were drawn. Jerne felt it was important to obtain Dutch input on including the Germans; however, the Dutchman Neurdenburg was in effect opposed (see Section 6.1). Ultimately, both of these options were dropped for varied reasons. Jerne was very thorough and solicited input from local members and duly conveyed these opinions to Bliss. It is recalled that concurrent discussions were going on with members in France, Belgium, Italy, and the Netherlands about other possible regional unions. Cutting across all the discussions were concerns about the handling of dues in different currencies. This was in addition to issues surrounding currency values for the day, with suggestions that some of these hardships could be allayed by establishing associate members who could 'share' the journal *Biometrics* among themselves. In July 1952, the picture became clearer vis-a-vis non-Scandinavian countries when Bliss advised Jerne that the Dutch group was not interested in joining with them, and that a separate Benelux group was being formed. However, Bliss still hoped a Scandinavian region could be formed; indeed, it was, but thirty years later. Meantime, Jerne's inquiries locally showed a real interest in members being willing to serve as national secretaries for national groups.

Jerne was based in Copenhagen and so, during his term on Council he was the de facto link between the international body through the Council and the local Danish members. Subsequently, Niels F Gjeddebaek was appointed as the National Secretary upon the formation of the Danish Group (GDe), serving from 1951 to 1980.

In 1953, Bliss sought advice from Herman O A Wold on possible national secretaries for Sweden; Wold suggested Harold Cramér who demurred saying the appointment should be Wold. Thus, Wold accepted the appointment in November 1953, with the formation of the Swedish National Group (GSD).

Group Norway (GNo) traced its formation back to 1960 when Secretary Healy contacted members indicating there were enough members to form a national group; in particular, he appointed Lars Strand in July 1960 to become the inaugural National Secretary, a position Strand held until 1970.

The desire for a region continued however. Strand in 1967 recognized that the Scandinavian groups were now "well established" but still thought a region could be formed and that it should include Finland. Cox expressed such a hope in 1969 when contacting the Finish biostatistician Erik Gustav Elfving about potential members in Finland. In contrast, in a February 1970 letter to President Berthold Schneider and Vice-President Cox, Secretary Hanspeter Thöni reported that Danish National Secretary Gjeddebaek was opposed to a region because he felt the respective countries were too far apart geographically making it difficult to have joint meetings, and that the present arrangements with local meetings would still be needed.

The defining initiative to unite the three national groups along with at-large members from Finland and Iceland arose from informal meetings between members attending a Nordic Conference on Mathematical Statistics in Mariehamm Finland in May 1980, prompted by Swedish National Secretary Paul Seeger. A working group contacted (the eighty-five) members in all five countries. With the necessary encouragement to proceed, the national secretaries and a representative from Finland and Iceland contacted President Richard Cormack complete with Regional By-laws for the proposed region. Council approved, and so in early 1982, the new NR was a reality; Aage Vølund, a pharmaceutical biostatistician, was the first Regional President (see Figure 6.5). The By-laws defined a Regional Committee comprised of a Regional President (who could not serve for more than two consecutive years), regional secretary and regional treasurer, and up to five ordinary members to ensure each country was represented elected annually.

FIGURE 6.5
Aage Vølund (– 2021).

6.5 Hungarian Region (HR)/Hungarian Group (GHu)

The Society leadership was interested in promoting biometry and the Society in Eastern Europe, including Hungary. By 1963, Bliss was contacting Ireneusz Juvancz about possible activities in Hungary, including having the Society provide funds for a Symposium in 1967 organized by Juvancz. Thus, it was that the Hungarian Group (GHu) was formed in May 1965 with Juvancz appointed National Secretary; all members were also members of the Biometric Section of the Hungarian Biological Society, a double track similar to that in Poland. Juvancz seemed to be an energetic and enthusiastic member, achieving much in the way of organizing meetings and initiatives; even reporting to Secretary Le Roy "we do in little what the IUBS [International Union of Biological Sciences] does in great," having earlier told Bliss that "Sometimes indirect routes are quicker." [!] Juvancz continued as National Secretary until stepping down in 1980 for János Sváb; Sváb served until dying in office in May 1986 where-upon Elisabeth Baráth took over the post. Later, on October 4, 1988, the group became the Hungarian Region with Baráth as the first Regional President; see Figure 6.6. Baráth later served as Society Secretary (1993–2000) and was a key player in the invitation from the region to hold the 1990 IBC in Budapest. With the fall of the Berlin Wall only months before the Conference, members flocked to this IBC which was consequently a huge success (see Chapter 9, Section 9.1.15).

Currency issues dominated dues payment in Eastern European countries, Hungary was not exempt. Members here were quite creative in offering ways

FIGURE 6.6
Elisabeth Baráth.

to ease this burden, even suggesting at one stage that members be "paired up" with one payment in alternating years. For its IBC, the grant paid by the Society would be taken out of dues being held by the Hungarian Region but not forwarded to the Society.

The Region was dissolved in 2003, and reverted to National Group (GHu) status. As of 2011, it is currently inactive.

6.6 Spanish Region (REsp)

The origins of the eventual formation of a region in Spain trace to conversations between French Regional President Richard Tomassone and Carmen Santisteban Requena in early 1981. Tomassone quickly (in June 1981) informed President Richard Cormack of the Spanish interest. Finally, in May 1984 there were twelve members, sufficient to form a national group. Therefore, President Pierre Dagnelie reported to Council in April 1985 that, with more than twenty-five members in Spain, consideration should be given to the formation of a new national group. Thus, the Spanish National Group (GSp) was formed officially in 1985 with Carmen Santisteban Requena as the National Secretary. The Group rapidly gained membership until, with over eighty-seven active and thirty-seven inactive members, it became the Spanish Region (REsp) in 1992; Santisteban (see Figure 6.7) became the first Regional President. The region was quite active in promoting biometry, especially supportive in the creation of new regions and participated extensively in the Central American and Caribbean Region's network meeting in Venezuela. From May1985 until June 1992, the members published a newsletter *Biometria*

FIGURE 6.7
Carmen Santisteban Requena.

every two months which kept the membership informed of activities within the region/group as well as Society news. The Region hosted the 2018 IBC in Barcelona Spain.

In her summary of regional growth and its related efforts for the Society over the years, Santisteban concluded "Objetivos cumplidos, misión cumplida y momento de irse a la retaguardia." ("Objectives accomplished, mission accomplished and time to go to the rear.") In a real sense, though written to the members of this Spanish Region, the final region formed by the Society in its first fifty years, this is a somewhat apt descriptor of where the Society itself was positioned as its own first fifty years drew to a close.

6.7 Eastern Europe

Establishing regions or national groups in Eastern European countries was always a goal of Bliss. Indeed, his Grand World Tour (see Chapter 3, Table 3.1, for details) included stops first to Poland and then to the Union of Soviet Socialist Republics (USSR) (in April 8–14, and April 14–23, 1962, respectively) to generate interest in biometry and the Society. Currency and political issues were always important factors in any successful endeavor here.

Currency issues were essentially two-fold. One was the actual fee charged by banks to transfer monies to (in this case) the US to pay annual dues; the other was the amount of those dues for members in countries with relatively lower salaries. [Recall that at some stage, national groups or regions, could resolve the first problem, by collecting dues locally and sending one check to the Society covering all local members. Individual members naturally did not have that option.] Prompted by a question from a Russian member, payment for a five-year period was explored. Also considered was the idea that a member and/or region from elsewhere could sponsor someone from Eastern Europe (but this raised questions about foreign exchange regulations with some western banking laws) although there were recorded cases of a Western European member paying dues for an Eastern European At-Large member. As President David Finney opined in 1965, "there is willingness but ... difficulties" in resolving such ideas.

Since by far the major share of the dues was to cover the cost of *Biometrics*, the second aspect was harder to resolve. Through 1963 and 1964 there was a lot of discussion between the leadership (primarily Bliss, Finney, Tucker, Le Roy) exploring ideas as to how to make the journal accessible to Eastern European members at a reasonable price. In an interesting aside, Bliss said that the "soviets tend to pay promptly." One idea was to reproduce galley proofs (but that raised copyright questions). Another idea was to provide microfilms or microcards of the journals (but this could interfere with sales of back issues, a source of valuable income to the Society at the time). Still another idea involved photo-reproduction (but that proved to be too expensive). Frank Anscombe (BR) wondered if *Biometrics* abstracts could be published in Russian (which prompted the idea to publish abstracts in French and German also). Though proffered solutions were explored before being abandoned, the concept of finding an economic solution remained for further exploration. This included the possibility of an associated national grouping. Twenty years later, in his April 1985 Report to Council, President Pierre Dagnelie raised the question of how to maintain membership in these countries under the prevailing fee structures.

Throughout 1963, there were several letters between Bliss and different Soviet members discussing the formation of a Soviet Region. The aforementioned questions of how to make the journal accessible and in a cost-effective manner were constants. So far, no all-encompassing Soviet Group emerged, but separate Romanian and Polish national groups did become realities. Hungary, also part of Eastern Europe became a Region in 1988; see Section 6.5.

Meantime, as there were eleven members in Romania by 1968, the Society was anxious to form a national group there. However, currency difficulties persisted in all of Eastern Europe and so posed a problem still to be resolved. Secretary Hanspeter Thöni and Treasurer Henry Tucker (in 1970) looked into the possibility of the Society receiving one payment from the region to alleviate these difficulties. Another complication was that in much of Eastern Europe, formation of such groups was not allowed without contacting certain government officials. Meanwhile, with appropriate supporting letters from the Society, Romanian Tiberiu Postelnicu was active in (the not-inconsiderable) efforts to gain permission from the relevant government authorities to form a national group. Eventually, the Romanian National Group (GRo) was authorized in 1971 with Postelnicu as National Secretary. Later, in 1972, local member P Stefan Niculescu started soliciting interest in hosting an IBC in Romania; these efforts came to fruition with the Eighth IBC being held in Constanza Romania in August 1974 (see Chapter 9, Section 9.1.8)).

Cutting across these decades, biometric activity flourished in Poland. The first Pole to become a member was Julian Perkal (sponsored by WNAR members Jerzy Neyman and Elizabeth Scott); Perkal was quite active in recruiting potential members while Neyman continued to encourage biometry in the country, so much so that by 1960 Secretary Healy is hoping to have sufficient numbers to form a national group. Concurrently, a Polish Biometrical Society (PBS) was established in 1959 with 100 members, now numbered close to 200 in the late 1960s. Many PBS members were members of the Society; but for a variety of reasons, from the outset, attempts to form a national group did not succeed. The year 1980 saw a slew of letters between President Richard Cormack, Secretary Lyle Calvin and Treasurer Jonas Ellenberg as they explored ways to find a path through the local currency and political issues and the Society's constitutional boundaries; success eluded them. However, after talks with Society officers (President Richard Tomassone, Secretary Roger Mead and Treasurer Janet Wittes) while at the International Society for Clinical Biostatisticians (ISCB) meeting in 1991, a national group was formed in 1992 with Anna Bartkowiak as National Secretary. While the PBS and the Polish National Group (GPol) were independent organizations, most members of the National Group were also members of PBS, and both maintained contact with each other, with officers of one being officers of the other for example. Since 1964, the PBS has published its own journal *Biometrical Letters*, twice-yearly.

6.8 National Groups and Networks

Over time, regions were formed when the number of members in a specific geographical location reached a loosely identified certain number. In the early years, there was no specified number, while in later years this was about fifty. With fewer members, defined by 1952 as approximately ten-fifty, members would be coalesced into national groups. Typically, if a prospective member lived in a geographical location which already had a region or a national group, then he/she automatically became a member of that region or national group. Otherwise, the member would be called an "At-Large" member. When an At-Large person showed particular interest and energy, s/he was encouraged to form a national group. The Society would then appoint a National Secretary (often that same energetic member). Thus it was that many regions started out as national groups. Their developments have been described in the relevant chapters/subsections thus far. One anomaly is India. Indian members started out as a region prescribed at the Second Council immediately following Woods Hole. It formally came into existence in 1949, members converted to At-Large status in 1951, before becoming a National Group (GInd) in 1953, and returned to its regional status in 1989; see Chapter 5, Section 5.1.

In the mid-1980s, President Dagnelie and Treasurer Ellenberg developed a Manifesto "Towards (Greater) Internationalization of the Biometric Society" (see Dagnelie, 1986, and Ellenberg, 1988) as a forum for enlarging the geographical influence of the Society (shades of Bliss, no less). Extending the interactions worldwide among members and the strengthening of already existing but struggling regions and national groups were included in this endeavor. As a result, many new members and hence national groups emerged in new geographical parts of the world; already existing regions were encouraged to provide support to possible members in these emerging areas. Some national groups had already come into being, others came as a result of this initiative. Those national groups not covered thus far are considered here, each in turn.

Though the Dagnelie-Ellenberg Manifesto was somewhat formal, it was not the first such effort. Bliss had traveled the world in his "Grand World Tour" during his 1961–1962 sabbatical year. By building on Bliss' contacts, in the late 1960s, the September 1969 meeting of the Executive Committee declared its intentions to expand and form national groups in Socialist (Eastern European) countries, South America, and the East and Far East countries, with Le Roy, Tucker and Cox, respectively, to spearhead efforts in these countries; Cox would also reach out to the Scandinavian National Groups. In 1987, Secretary Mead tells Treasurer Ellenberg that he felt that the major Society need was the "development of new geographical groups and regions and the stimulation and encouragement ... in apparently non-active Regions."

Developments in Mexico ebbed and flowed over the years. Through the 1960s, there had been quite an effort spearheaded by Ted Bancroft (ENAR) along with Henry Tucker (WNAR) to support an international Symposium with the hope that this would stimulate interest in the Society; but unfortunately this symposium never transpired. Earlier, it had been determined, in its new 1960 By-laws that Mexican members would be part of the WNAR. However, Bliss (January 1965) thought it would be better for the then-four Mexican members to be At-Large members because of cost. Eventually, in April 1967, the now twenty At-Large members present voted unanimously to form a national group (Group Mexico, GMex) with Eduardo Casas Díaz as National Secretary, duly approved in May 1967.

In the late 1960s, Cox has been working on the development of biometric interests through her contacts in Indonesia. President Schneider opined that her efforts were 'bearing fruit', to the tune that Andi Hakim Nasoetian was requesting a national group be formed. This National Group Indonesia (NGi) was approved by Council in late 1971, effective from January 1972, with Nasoetian as the first National Secretary. The Group became dormant in 1991, but there had been no request for its dissolution even though the membership had dropped to zero. The Group initiated its reactivation in 1995.

As early as January 1972, Brazil's Pimentel Gomes had taken the initiative to start a national group in nearby Argentina with unanimous support from the Brazilian Region. Argentina's María Esther Suárez was enthusiastic. Letters flowed between Gomes, Suárez, Brazil's Regional Secretary Aldir Alves Teixeira and Secretary Thöni; Suárez was to be appointed as National Secretary in 1972. However, the impetus seemed to stall in the mid-1970s, with members still registering as At-Large members. Elsa C Servy (in March 1975) wanted to try again especially since many members of the Statistical Society of Argentina founded in 1952 wanted to be part of the Society; and so Thöni replied with details of what was needed to establish their group constitution and by-laws. The president of the Statistical Society of Argentina returned to the question in December 1995 voicing the interest of some of his society's members wanting to formalize relations with the Society, though he is still wary of potential difficulties of split loyalties; analogies with other like-situations (such as the Biometrics Section of ASA and ENAR) helped to assure him that both components could indeed co-exist. Ultimately then in 1996, Argentina became its own National Group (GArg), with R B Ronceros as the first National Secretary, and became a Region (RArg) in 1999 with Nélida Winzer; as the first Regional President.

Dagnelie-Ellenberg's Manifesto was clearly heard and discussed by the regions, even as it was still being re-drafted before its publication in the 1986 *Biometric Bulletin*, spawning a flurry of new activity. For example, by 1986, the Belgian Region was sponsoring members in Morocco and Benin, while the British Region taking its lead from Belgium sought to sponsor members in Kenya, Malawi and Zambia, and in Central and East Africa more generally (more below). A group formed in South Korea (GKo) in 1988, with Sung H Park as the first National Secretary. Earlier (1985), ENAR had decided to sponsor Chinese members as well as library subscriptions in China and Romania.

A Chinese National Group (GCh) started in 1987, with Ji-Qian Fang as the first National Secretary. In an interesting twist, during the 1980s, there had been innovative approaches to the Society from Chinese personnel, e.g., one to publish the journal and another to purchase and distribute computer software (through a "software house"), in both instances intending to use the proceeds to cover the dues of Chinese members. Initially sympathetic, these unusual requests were considered by the Society at length. In the end, the offers were declined, for a variety of reasons, including that no one Region could be viewed as being "more equal than others" (recall the Indian Region dues problem, in Chapter 5, Section 5.1), and that individuals/regions could not make decisions that have policy implications, nor according to the Constitution could any member receive any benefit from nett earnings or operations of the Society. After lapsing in 2009, the China Group was reinstated in 2012.

There was considerable activity on the African continent. In 1989, a National Group formed in Kenya (GKe) supported by the British Region with Manibhai S Patel appointed as Acting National Secretary, as was a national group in 1989 in Morocco (GMo) with support from the Belgian Region and Dagnelie with National Secretary Ahmed Goumari. A South African Group (GSAf) which later hosted the 1998 IBC started in November 1990 becoming an official National Group in 1993, with Arthur Asquith Rayner as National Secretary. This South African effort was quickly followed by the Zimbabwe Group (GZim) in 1991 with Sr Jane Canhão as the National Secretary. Later, in 1994, the Botswanna Group (GBot) emerged (when Patel relocated here and became its National Secretary). To these groups, Nigeria (GNi, formed in 1997, Sagary Nokoe the first National Secretary) and Uganda (GUg, formed in 1994 and revived in 2010, with Leonard K Atuhaive as the first National Secretary) were added. The members through the various national groups were most energetic and would often gather for scientific meetings as loosely defined networks. In 1989, we learn of a Central African Network; a second East Central and Southern African Network meeting took place in Harare Zimbabwe November 25–29, 1991; the third meeting of the East Central and Southern African Network was held in Kampala Uganda in 1993 (with funding support from the NR); an Eastern and Southern African Network existed in 1994; and then later in 1995 there was a fourth meeting of the East Central and Southern African Network at Stellenbosch South Africa. Throughout the

African Network Participants in Search of Source of the Nile

FIGURE 6.8
Participants at Eastern and Southern African Network Meeting, Kampala Uganda, September 13–17, 1993, in search of data, i.e., the source of the Nile River; Erica Keogh (GZim), Tim Dunne (GSAf, Agnes Ssekiboobo (GUgan), -, Sr Jane Canhão (GZim), -,-, Rob Kempton (BR).

African continent, during the 1980s, first initiated by Secretary Roger Mead and *Biometric Bulletin* Editor Joe Perry and consolidated by Rob Kempton (all British Region members) worked tirelessly encouraging the formation of these various national groups and networks; many of those members who worked toward these endeavors are shown in Figure 6.8.

A Caribbean Network, spearheaded by former Treasurer Larry Nelson and then-current Secretary Mead, was being established in the early-mid-1990s. The first meeting of the Central American Caribbean network was held in July 1992. By 1995, a Network consisting of members from the Caribbean, Central America, Columbia and Venezuela was emerging, forming as a result of scientific meetings. A new national group was formed in Guatemala (GGuat) in 1995; Jorge Matute served as the first National Secretary. Group Columbia (GCol) formed in 1995 with P Pacheco as National Secretary.

A meeting of regional presidents and national secretaries at the 1994 IBC concluded that, rather than a network becoming a region, a network was an interim stage and that each of its component national groups should be encouraged to become a region eventually. The SUSAN network (1999) and the Caribbean network (2002) showed the way; see Chapter 11.

Over the years, other initiatives to form specific national groups emerged; e.g., Sarhan approached Cox in 1972 to form an Egyptian group, and there was an outreach to Bangladash with its more that ten members in 1996. So far however, these efforts had not come to fruition.

In contrast to regions which are governed internally by elected regional officers, with terms of office varying from office to office and across regions, national groups on the other hand liaised with the Society through a National Secretary appointed by the President though usually in consultation with members of the proposed national group; there is no term limit on National Secretary appointments.

7

Publications – Biometrics

7.1 Introduction

The scientific backbone of the Society is its flagship journal *Biometrics*. While its scientific importance is paramount, it is also the backbone of the financial structure of the Society; indeed, deliberations about dues structures revolved around the cost of this journal.

The *Biometrics Bulletin* first appeared in 1945 as a bi-monthly newsletter published by the Biometrics Section ("Section") of the American Statistical Association (ASA) designed to keep its members informed of events since ASA was unable to meet due to World War II; see Figure 7.1. [This publication should not be confused with the similar-but-differently named *Biometric Bulletin* newsletter which was first published in 1984; see Chapter 8, Section 8.1.] Chester Bliss as Section Chair, appointed Gertrude Cox as the first editor. The 1945 ASA President Walter Shewhart acknowledged this new undertaking with "The launching of the Biometrics Bulletin is a logical step not only in fostering contacts between biologists concerned with statistical information, problems, and methods but also in stimulating research and in elevating the standards of statistical work that should prove helpful in developing the profession of statistics." The year 1947 saw a change to a quarterly publication and a name change to *Biometrics* for the first issue (March 1947). The latter action followed Bliss' discovery that a Dr Jellinek had started a journal in 1936 with the name *Biometrics Bulletin* but it was discontinued after one year (of four issues); consequently, Bliss bought the name from Jellinek for $1 and changed the name, see Fertig (1984).

Initially, this was an ASA journal; thus, as such, ASA controlled policies and management questions. Unsurprisingly, by 1948, the Society wanted its own journal and in particular thought that *Biometrics* would meet its needs. We describe the process of transferring *Biometrics* from ASA to Society ownership first (Section 7.2). Then, the content of *Biometrics* is discussed in Section 7.3, a coverage of editors and editorial policy changes over the years is presented in Section 7.4, and how the journal production was managed is considered in Section 7.5, concluding with a brief look at the overall importance of this journal (Section 7.6). An overview of changes, not mentioned in this chapter, in the publication process since 1997, is summarized in Chapter 11. Finally, for the record, Chapter 12, Section 12.3 provides a

DOI: 10.1201/9781003285366-7

February 1945

BIOMETRICS

Vol. I No. 1 BULLETIN

THE BIOMETRICS SECTION, AMERICAN STATISTICAL ASSOCIATION

On The New Biometrics Bulletin

When the American Statistical Association was founded more than a hundred years ago, statisticians were primarily concerned with the collection of data of interest to the State. Within the last sixty or eighty years, however, statisticians have developed methods that have entered one branch of science and industry after another and have attained a central position in biology, physics, chemistry, meteorology, and astronomy, as well as in the social and economic sciences and in mass production and distribution.

The American Statistical Association has not adequately fulfilled its responsibility with respect to this development although its purpose is stated in part "to foster contacts among persons seriously concerned with statistical information, problems, and methods, . . . and to encourage the application of statistical science to practical affairs."

The formation of the Biometrics Section within the American Statistical Association in 1938 marked an important step toward correcting this situation. This step recognized and gave support to the work of biometricians, who since the time of Galton have made important contributions to the subject matter of their own field and have done much to develop the modern statistical method that is of equal importance in many other fields.

The launching of the Biometrics Bulletin is a logical step not only in fostering contacts between biologists concerned with statistical information, problems, and methods but also in stimulating research and in elevating the standards of statistical work that should prove helpful in developing the profession of statistics.

WALTER A. SHEWHART, *President.*
American Statistical Association.

Biologists and Biometricians

With this first issue of the BIOMETRICS BULLETIN, the Biometrics Section addresses a larger audience than in the past. Biologists and biometricians form a constructive symbiotic pair and the Section is primarily concerned with furthering this relation. Many of our members are now absent on war assignments and we have few opportunities to interchange ideas at national meetings. Under these conditions the BIOMETRICS BULLETIN will be invaluable as a means of communication and as a medium for bringing our biological friends into a coherent organization. The high calibre of our Editorial Committee insures a publication which the biologist can trust. We are especially fortunate in having as the Chairman of our Editorial Committee, Professor Gertrude M. Cox, Director of the Institute of Statistics at North Carolina State College.

C. I. BLISS, *Chairman,*
Biometrics Section.

FIGURE 7.1
First issue of *Biometrics Bulletin* February 1945.

list of those who served the Society as Editors of *Biometrics*. Regional and national group abbreviations are provided in Chapter 12, Table 12.1.

7.2 Transfer of *Biometrics* from ASA to the Society

At the Society's formation, Cox offered *Biometrics* as an outlet for Society news including the publication of the Woods Hole 1947 Proceedings. While this was helpful to the nascent society and although the Society paid for the space used (in 1949, the charges were $20 per page), the journal belonged to ASA and as such the Society had no say in its management.

As previously indicated, by 1948, the Society felt it needed its own journal, and in particular, it seemed that the Society should take over *Biometrics*.

Thus, discussions began with ASA President George Waddel Snedecor and with the Biometrics Section during the annual meetings in December 1948. Snedecor felt it was logical that the journal "follow the Biometric Society rather than the Statistical Association" and did not rule out the possibility that it becomes a Society journal. At the parallel Eastern North American Region (ENAR) Regional meeting, two options were outlined: one that the journal continues as an ASA journal, or two that the journal becomes a Society journal. If the journal stayed with ASA, then under the ASA's new constitution, control would be through the ASA Publication Committee and so the Society input would be limited to any provided by ENAR members who were also Section members. All indications pointed to the fact that no one wanted the journal to stay with ASA. Accordingly, the Society's Council in February 1949 formally voted to enter negotiations to take over responsibility for *Biometrics* and to appoint a special committee to effect this. The ASA's new constitution implied that if this were to happen, then January 1950 should be the target operative date.

Meantime, through 1949, discussions were occurring within ASA and with its Section members. Samuel Stanley Wilks (1949 ASA President-elect) was taking an active lead role. It seemed to ASA that either the Society would take over publication of *Biometrics* or start its own journal. If the former, some discussions included that there should be some statement in journal issues referring to its having been "founded" by the Biometrics Section (this happened until the practice was discontinued with the 1968 issue); if the latter, then there were concerns for the future viability of *Biometrics* and even of the Section itself. Would Section members continue to provide the manpower behind the production of an ASA controlled *Biometrics* or would those energies be transferred to any new Society journal? Also, the continued existence of the Section seemed to some a very legitimate concern since all its members received *Biometrics* at the reduced rate offered to ASA members. Bliss tried to give assurances that the Section should continue to exist especially in its scientific programmatic roles for ASA meetings. [Interestingly, analogies were being drawn with the arrangements made when the Institute of Mathematical Statistics (IMS) took over control of the *Annals of Mathematical Statistics*.] These discussions heightened the realization that this transfer was a real possibility, to the degree that some ASA members were starting to evidence signs of "cold feet." Some components of ASA were opposed because of a perception that the Society and the journal had been formed prematurely. To that Snedecor (now ASA Past President) proclaimed to Bliss in December 1949, "[this action] ... was premature. But, after all, it doesn't so much matter when a baby is born; it is its subsequent growth and well-being that determine its usefulness. Don't bother about the past, but look to the future." After a year of constant delays, surely Bliss was encouraged by this affirmation.

A concurrent issue here was that ASA had operated at a loss in 1948. One consequence was that ASA determined that editorial and financial responsibility of its journals was to be transferred to the editors in 1949; Cox felt that, although production/financial details were as yet not fully known to her, she should be able to manage this though it would probably be necessary to move journal operations from the ASA Washington office to her home base in Raleigh. It was also apparent that some increase in subscriptions/dues would be necessary. The Society had hopes that a plan to seek a subsidy from the United Nations Educational, Scientific and Cultural Organization (UNESCO) to cover conference proceedings costs would materialize, an action only possible if the Society owned the journal.

Thus, a June 1949 vote of Council approved (twenty-one to one), though perhaps understandably in an earlier vote two members wanted better assurances that the Society could indeed successfully navigate any potential financial problems. Accordingly, President Ronald Fisher appointed John Hopkins (Chair), Bliss and Cox as a Special Committee to work on the necessary details. By May 1949, the proposal was that ASA was to transfer *Biometrics* to the Society effective from the first issue of 1950, along with two stipulations. The first stipulation was that "ASA members would be entitled to a block subscription rate less than that given to outsiders under an arrangement reversing that by which the Society now sends *Biometrics* to its members ... for perhaps five years." (This rate was set at $3 for 1950.) The second stipulation was that back numbers up to 1948 would be handed over to the Society, and that half of the nett proceeds from subsequent sales would be paid to the ASA. In light of the concerns over finances, the Special Committee added a third stipulation that should publication of *Biometrics* be discontinued within five years after transfer, then ASA would "have the right of recovering without charge title ... and all undistributed back issues."

By June 1949, Bliss (as only a pure Bliss could be) was becoming impatient with an apparent slowness (foot-dragging?) from the ASA and its Section Chair (Margaret Merrell) with what Hopkins "suspect[ed was] a certain amount of passive resistance in that quarter." Editor Cox too was frustrated with the state of affairs saying "I almost feel that much of this delay is deliberate." She appeals to Bliss to "break the deadlock." Bliss wanted more definitive proposals from ASA since, as he reported to them that the whole issue and its ramifications were to be considered at length at the upcoming Geneva International Biometric Conference (IBC, August 30 to September 2, 1949). Eventually, late but in time for the ASA's October 1949 Board meetings, Merrell produced a report outlining reasons as to why none of the proposals was acceptable to her; however, she ended by saying she was "not representing the views of a unanimous Committee" instead saying that the majority were in favor (of the proposals), and that divergent views had been reported to ASA officers.

Finally, the ASA Board of Directors ("Board") meeting of October 29, 1949 discussed the issue; Bliss was invited to be in attendance. Although the Board acknowledged that Section members had with a large majority in a recent summer ballot approved this transfer, the Board wanted the Section to discuss the question further and confirm this decision at its business meeting in

December (29th 1949). The Board also wanted the Section to ascertain its own future policy and whether or not the Section accepted ENAR's application for affiliation with the Section. An implicit Board concern here was the future existence of their Biometrics Section; subsequent events proved this to be an unnecessary concern, with both the Section and ENAR continuing to co-exist successfully still. If Section confirmation came forth at the December meeting – it did, unanimously – the Board would appoint a Negotiating Committee empowered to act so that the transfer would be completed by the time of the first issue in 1950.

The ASA's appointed members to its Negotiating Committee were its Biometric Section 1950 Chair (Harold F Dorn) and the ASA 1950 President-elect (Lowell Jacob Reed). The Society's Special Committee (Hopkins, Cox, Bliss) would become its Negotiating Committee. By April 1950, Reed had written agreeing to the Society's terms with but two minor modifications, one of which asked that the Section could nominate one member (approved by the Editor) to the Editorial Board. The second concerned associate members, a category of membership still be to defined, and so to be clarified later. On October 2, 1950, members were informed that *Biometrics* had become the property of the Society. See the Agreement between ASA and the Society in Figure 7.2.

That should have settled the matter. However, on October 25, 1950, Bliss took a phone call from Samuel Weiss (ASA's part time first executive director), who conveyed the news that at the just completed ASA October 1950 Board meeting, it was decided to withhold $0.50 of the block subscription cost of the journal to ASA members for handling charges. Not surprisingly, Bliss and his Negotiating Committee co-members were totally taken aback, since as Bliss tried to explain to Weiss this amounted to a rebate which was not part of the Agreement. Weiss wanted an acceptance of these terms immediately as he had to send out dues notices. There was a back-and-forth about giving proper notice of pending actions to which Bliss pointed out that the original agreement intended to be signed in January for a March issue start was not completed until mid-August due to inaction on the part of ASA representatives. However, Weiss was not to be moved – he was in fact only implementing Board decisions – so Bliss, forced to make an on-the-spot decision, accepted that the Society would take the loss but for 1951 only, expressing his "disapproval" in "unmistakable terms."

The original block agreement for ASA members was to remain in effect for five years (i.e., until 1955); it was renewed in 1956. It had seemed like a good idea back in 1950, but by 1956-1957-1958, the subscription numbers suggested that there was a severe imbalance in favor of ASA and disadvantageous to the Society. Even ENAR's membership campaign failed – ENAR had sent 750 invitations to join the Society including all 547 ASA members who subscribed to *Biometrics* through the ASA block agreement. The upshot was that an estimated 650 ASA members planned to participate in this block arrangement in the following year (1957) but without also joining the Society. Those who

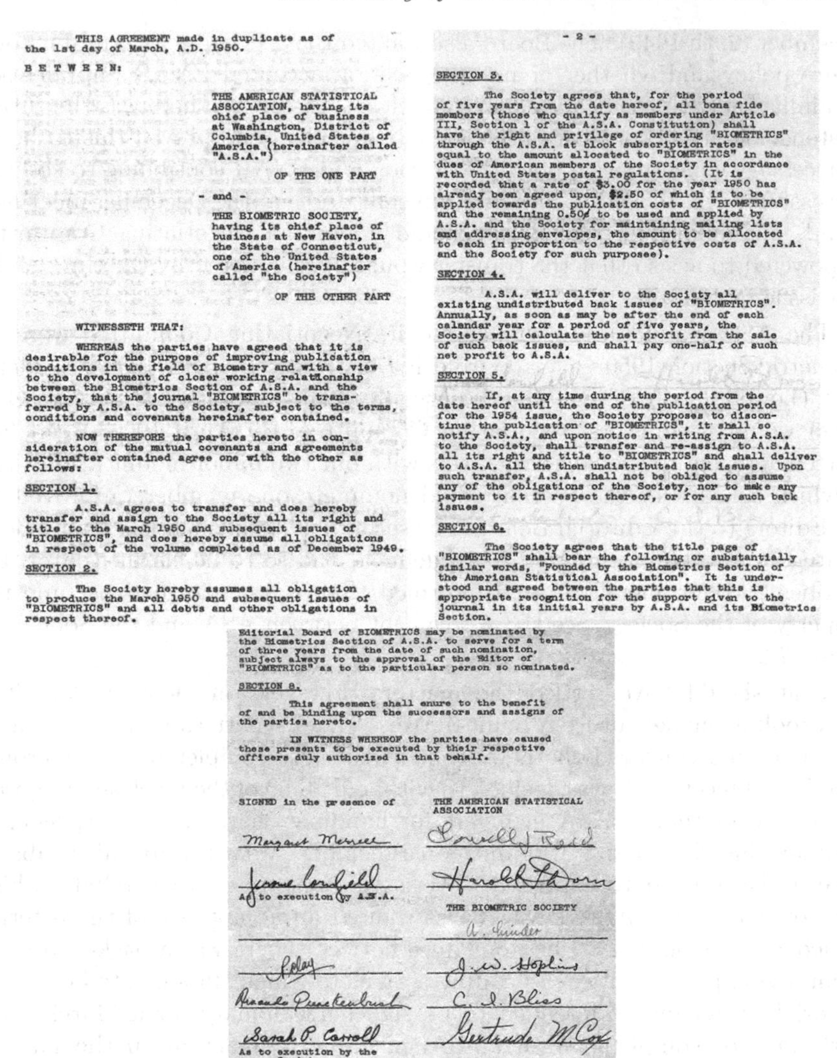

THIS AGREEMENT made in duplicate as of the 1st day of March, A.D. 1950.

B E T W E E N:

THE AMERICAN STATISTICAL ASSOCIATION, having its chief place of business at Washington, District of Columbia, United States of America (hereinafter called "A.S.A.")

OF THE ONE PART

- and -

THE BIOMETRIC SOCIETY, having its chief place of business at New Haven, in the State of Connecticut, one of the United States of America (hereinafter called "the Society")

OF THE OTHER PART

WITNESSETH THAT:

WHEREAS the parties have agreed that it is desirable for the purpose of improving publication conditions in the field of Biometry and with a view to the development of closer working relationship between the Biometrics Section of A.S.A. and the Society, that the journal "BIOMETRICS" be transferred by A.S.A. to the Society, subject to the terms, conditions and covenants hereinafter contained.

NOW THEREFORE the parties hereto in consideration of the mutual covenants and agreements hereinafter contained agree one with the other as follows:

SECTION 1.

A.S.A. agrees to transfer and does hereby transfer and assign to the Society all its right and title to the March 1950 and subsequent issues of "BIOMETRICS", and does hereby assume all obligations in respect of the volume completed as of December 1949.

SECTION 2.

The Society hereby assumes all obligation to produce the March 1950 and subsequent issues of "BIOMETRICS" and all debts and other obligations in respect thereof.

- 2 -

SECTION 3.

The Society agrees that, for the period of five years from the date hereof, all bona fide members (those who qualify as members under Article III, Section 1 of the A.S.A. Constitution) shall have the right and privilege of ordering "BIOMETRICS" through the A.S.A. at block subscription rates equal to the amount allocated to "BIOMETRICS" in the dues of American members of the Society in accordance with United States postal regulations. (It is recorded that a rate of $3.00 for the year 1950 has already been agreed upon, $2.50 of which is to be applied towards the publication costs of "BIOMETRICS" and the remaining 0.50¢ to be used and applied by A.S.A. and the Society for maintaining mailing lists and addressing envelopes, the amount to be allocated to each in proportion to the respective costs of A.S.A. and the Society for such purposee).

SECTION 4.

A.S.A. will deliver to the Society all existing undistributed back issues of "BIOMETRICS". Annually, as soon as may be after the end of each calendar year for a period of five years, the Society will calculate the net profit from the sale of such back issues, and shall pay one-half of such net profit to A.S.A.

SECTION 5.

If, at any time during the period from the date hereof until the end of the publication period for the 1954 issue, the Society proposes to discontinue the publication of "BIOMETRICS", it shall so notify A.S.A., and upon notice in writing from A.S.A. to the Society, shall transfer and re-assign to A.S.A. all its right and title to "BIOMETRICS" and shall deliver to A.S.A. all the then undistributed back issues. Upon such transfer, A.S.A. shall not be obliged to assume any of the obligations of the Society, nor to make any payment to it in respect thereof, or for any such back issues.

SECTION 6.

The Society agrees that the title page of "BIOMETRICS" shall bear the following or substantially similar words, "Founded by the Biometrics Section of the American Statistical Association". It is understood and agreed between the parties that this is appropriate recognition for the support given to the journal in its initial years by A.S.A. and its Biometrics Section.

Editorial Board of BIOMETRICS may be nominated by the Biometrics Section of A.S.A. to serve for a term of three years from the date of such nomination, subject always to the approval of the Editor of "BIOMETRICS" as to the particular person so nominated.

SECTION 8.

This agreement shall enure to the benefit of and be binding upon the successors and assigns of the parties hereto.

IN WITNESS WHEREOF the parties have caused these presents to be executed by their respective officers duly authorized in that behalf.

SIGNED in the presence of

THE AMERICAN STATISTICAL ASSOCIATION

AS to execution by A.S.A.

THE BIOMETRIC SOCIETY

As to execution by the Society

FIGURE 7.2
Agreement: Transfer ownership of *Biometrics* from ASA to the society.

said so frankly admitted that they used this arrangement to save $3, but were equally clear that they would gladly pay that extra $3 to join the Society to obtain the journal. A concurrent reciprocal arrangement allowing for Society members to receive the *Journal of the American Statistical Association (JASA)* at a block rate was used by but a scant 26 members (especially ENAR members) in 1957. The block subscription ($4) barely covered the basic cost (about $3.60 per year) of the journal excluding overheads. This amounted to a financial imbalance on Society members.

Thus, by 1958, there was general agreement to terminate this block privilege arrangement. Details of the gains and loses in members and commensurate income were carefully worked out, with a conclusion that if only 370 of the projected 650 for the year 1958 joined the Society, then the Society and the journal would break even. There were also some confusions along the way, occupying time and attention requiring clarifications to various folk; e.g., some who participated via ASA's block arrangement erroneously thought they were automatically a member of the Society, some joined 'both' organizations and received two copies of the journal, among other mostly irritating misunderstandings. It was assumed that since the arrangement with ASA was reciprocal, then the arrangement for Society members to receive *JASA* at a reduced rate would also terminate. Council was asked to approve this in April 1958. Later in the year, ASA's President Walter Hoadley contacted the Society's President Cyril Goulden with his concerns about "possible deep permanent cleavage" between the two associations and asked if the Society would reconsider. The Society did not reconsider nor did any permanent fissures occur.

7.3 Content

For its first two years, the *Biometrics Bulletin* operated as a bi-monthly newsletter, its role to keep ASA's Biometric Section members informed about activities and events of professional interest, though each issue typically contained one scientific paper. During its second year, it was determined that the newsletter's content be expanded to include more articles with scientific content. Thus, the March 1947 issue started with a three paper set, Eisenhart (1947), Cochran (1947), and Bartlett (1947), which transformed the journal from a struggling forum to a major scientific outlet.

The three Eisenhart-Cochran-Bartlett papers looked at crucial aspects of experimental designs. The bar was set high. To these three are added Cochran (1954a,b); the first of these, Cochran (1954a), presaged later works on meta-analyses. Further, Cochran (1955) is a classic on vaccine trials. In the first decades, articles were almost exclusively focused on statistical methodology, especially aspects of regression and experimental design, used extensively in biology and agriculture and related fields such as forestry, fisheries, and plant sciences (both area and technique broadly defined). Articles were driven by applications in crop and plant sciences, poultry and animal sciences, plant and animal breeding, species selection, with genetics per se largely non-existent though there was a paper on the fruit fly *Drosophila* in the first issue and most of the genetics papers that did appear dealt with animal breeding and crop sciences. There were but few papers on medically related methods. A summary of content topics for the first twenty-five years is found in Cox (1972).

This early trend however evolved until by the 1990s, in the journal's fifth decade, this was reversed with a dominance of medical articles especially clinical trials and very few motivated by agricultural applications. The emergence and then explosion of statistical genetics in its myriad of manifestations drove this change in focus. In a different direction, the first decades also saw many articles on how to persuade the calculators of the day (including pen and pencil, and calculating machines) to produce useful calculations for the applications' analyses. With the advent of modern computers starting in the 1960s, this category of subject matter phased out to be replaced by new ideas only possible and accessible with these 'new' modes of computation; thus, again by the 1990s, there were papers on bootstrap, Gibbs sampling, imaging (fMRI) and the like. A Jubilee Issue in 1972 was devoted to multivariate analysis (edited by Australasian member Evan J Williams).

Regardless of application or methodology, there was an ever persistent thread of a tension between a too-much-mathematical focus and a not-enough-mathematical focus in journal articles; these contrasts were sometimes reflected in Editor Reports, some came as letters from members to the Society. Bliss (1958) in his account of the first decade of the Society addressed this growing and pervasive perception that our journal had become too mathematical with "One [of] the periodic complaint[s] of our biological members [is] that the journal is becoming too 'high brow' statistically for them to understand, and the counter-complaint of the Editor that good biological, less technical manuscripts are hard to come by, despite numerous pleas for material. After all, editors cannot accept papers that are not submitted for publication." Bliss did go on to admit that "over the years [*Biometrics*] has become more technical and advanced statistically." A 1953 proposal from the Belgian Region concerned, again from Bliss (1958), "a second journal to be called *Acta Biometrica* with more emphasis on quantitative biology and less theoretical statistics, ... , and to draw primarily upon papers given at regional meetings in continental Europe." The proposal was discussed at length by members and Council, "[but they] were unable to solve essential details such as its editorial policy, relation to *Biometrics*, and financing, so that the proposal was never implemented." Indeed, in 1954, Council had approved a committee (Finney, Geppert, Tage Kemp, Martin, van der Laan, and Vessereau) be set up to explore this possibility. There was a wide diversity of opinions though a majority in favor, but outreach to a number of possible funding foundations was not successful. Ultimately, this journal was subsequently produced by the Belgian Region. The issue did not disappear however, receiving much attention by the British Region in 1961, but also letters revealed a "growing uneasiness ... in France, Germany, Austria and Switzerland." Fifty years after the launch of *Biometrics*, a more applications oriented and less mathematically focused new journal, *Journal of Agricultural, Biological, and Environmental Statistics (JABES)*, was launched; see Chapter 8, Section 8.2.

The so-called newsletter aspects were continued throughout until the publication of the new outlet *Biometric Bulletin* in 1984 took over these roles (see Chapter 8, Section 8.1). Thus, notices of Society activities were retained. This covered the gamut of news about past and forthcoming meetings – IBCs, Symposia both regional and international, regional meetings, including the abstracts of presentations at these meetings – financial reports of both *Biometrics* and the Secretary's office, Council actions, Regional By-laws and their revisions, regional reports which reports might convey news about regional and national group activities including regional election results, as well as news about the prospective formation of new national groups and new regions, or new committees, and so on. In short, it really was the Society newsletter. There were also news items of members; indeed, for someone so inclined, it is possible to track movements of members from one position/institution to another. This latter role was clearly appreciated given its persistence over the years. Forty years later (*Biometric Bulletin* May 1988, p. 19) in a retrospective, David Finney opined that the March 1947 issue was particularly good but that he would like the "column of social gossip be abandoned." [Pure Finney!] As an aside, if Finney had one characteristic, it was his consistency; forty years earlier, in June 1948, on assessing the development of *Biometrics*, he thought "the journal is very well worthwhile ... The only suggestion ... is that the column of social gossip be abandoned." Obviously, that social news stayed.

A section on "Queries" under Snedecor was started in the first issue whereby questions were posed and answered by experts (which read like a list of "who's who" of our early pioneers). This was expanded in 1957 to "Queries and Notes" to encourage short submissions (up to 1500 words) dealing with methodologies related to applications work. After fourteen years of stalwart service, Snedecor was succeeded in 1959 by Finney who along with successor subeditors assumed a more regular three-four-five-year term. This section was replaced by a newly named Shorter Communications section in 1974; the Queries and Notes Editor C David Kemp at the time continued as the first Shorter Communications Editor.

In 1960, the concept of an Index emerged. Horace W Norton (Subject Index Special Editor, later Associate Editor for the Index, ENAR) was appointed, and so began the task going back to the first volume. After indexing the first twenty-five years of *Biometrics*, Horton asked to be relieved of the task; he had become an "institution" and his work for the 1945–1970 volumes was greatly appreciated. His successor was Miles Davis (1971-, ENAR), but there is no more information in the Archives until 1989 when the Finance Committee approved that the volume containing the indexes for Volumes 21–40 not be published, deemed to be too expensive and no longer needed. However, the *Current Index of Statistics Extended Database* (CIS/ED) contained bibliographic information for *Biometrics* records for 1965–1992. The issue re-surfaced in 1996 when the Editorial Advisory Committee debated whether or not to produce a printed version to complement CIS which had indexed *Biometrics* since 1975 in electronic form.

The idea of having a Book Review section was discussed by the Editorial Board in 1960, having been broached by Editor Ralph A Bradley in his

November 1958 Report to the Editorial Board. John G Skellam (BR) was the first Book Review Editor; the first review appeared in the June 1960 issue.

In his 1961 Report to the Editorial Board, Bradley asked for suggestions to improve the journal; one such proposal was that there be "invited papers" (so designated in the resulting articles) designed to be expository presentations of "important material in an authoritative way" but sufficiently elementary for profitable consumption for the applied medical and biological reader. It was felt that this innovation should help diffuse the difficulties round the debates between too mathematical and not enough application-oriented journal articles. Eventually, an Associate Editor was appointed to deal with "Expository Papers," later referred to as "Invited Papers"; the first such person was S Clifford Pearce (BR). Marvin Schneiderman (Invited Papers Editor) in 1970 addressed the difficulty of obtaining these necessarily long and detailed papers and wondered if it would not be preferable to seek shorter review papers, an idea almost unanimously supported by the Editorial Board and Associate Editors. A Consultants Column was started 1977.

Papers could be written in any of the three official languages – English, French, or German. Secretary Henri Le Roy was pleased to note that his contribution in German among the papers presented at the 1963 Cambridge IBC would be the first paper of "international character" in *Biometrics.* Later, ENAR proposed that a summary in some other language be provided for any paper. This was not approved however in a November 1965 Council vote. Nevertheless, under the leadership of Richard Tomassone (RF), French translations of paper abstracts started in 1967 appearing at the end of a paper; these were collectively moved to end of an issue in 2005. Translations were provided from Belgian Region and French Region members. The effort ceased with the last issue of 2009.

With the untimely death in 1956 of John Wishart ("drowned by misadventure"), Editor John Hopkins observed that *Biometrics* had no policy regarding formal obituaries. He suggested that the Society might be prepared to memorialize members of distinction, though he did not think this was for an editor to decide. Many years later, Council formulated a policy in 1982 during the Toulouse IBC. Obituaries began in the *Biometric Bulletin* in 1988, but its editor suggested that the Editorial Advisory Committee should establish guidelines to formalize whether any particular obituary should go into *Biometrics* or *Biometric Bulletin* or both and to what extent. The guidelines established in 1982 became the accepted procedure, viz., that a brief announcement of a member's death could appear in *Biometric Bulletin*, and that, for "Biometricians of International Status" and "Distinguished Statisticians with Biometric Interests," the President and Editor could initiate preparation of a substantial obituary for publication in *Biometrics*; and for those with national status, the regional officers could recommend a short (maximum one page) obituary for *Biometrics.* Further, there could be a memorial session at the earliest feasible IBC for those with international status, and regions were encouraged to hold memorial sessions at their own regional meetings.

7.4 Editors and Editorial Policy

Gertrude Cox was the founding editor. She wanted to step down effective from December 31, 1954, after a ten-year term. [She was the newly elected 1955 ASA President-elect.] Ultimately, she continued through to December 1955 as Editor, though new submissions in 1955 were handled by her successor (Hopkins) who served as an Associate Editor in 1955.

Attempts to select an editor from the United Kingdom were unfruitful. The Special Committee set up to select the next editor eventually recommended John W Hopkins (ENAR) as her immediate successor, the same Hopkins who was an integral part of the story when the journal was transferred to the Society; see Section 7.2. In a wonderful piece of mentoring and insightful advice, upon his acceptance of the role, Cox wrote to Hopkins with "I didn't like my own Journal work much in the first two years, because there are many decisions to be made and an editor always comes in for some complaints. But after a while this didn't bother me and I found the work quite educational." Indeed, as early as 1946, she was asking her Editorial Committee "When can I resign?" and again later in 1947, she wanted 'out' as depicted in Figure 7.3.

Hopkins was based at the National Research Council in Ottawa Canada, and officially took over in January 1956 for a five-year term. Soon he became quite ill, and work on the journal was severely impacted. Hopkins reported in May 1956 that he had "not completely recovered." By March 1957, he wrote to Secretary Michael Healy that owing to his "problematical" condition, "the Society machinery should be set in motion ... to nominate another Editor ... forthwith." Hopkins' assistant Florence Suddon was sending Healy manuscripts for him to handle but did assure him that the business end was up-to-date however. Hopkins' illness was so serious, that in April 1957 then-President Alf Cornish appointed a committee to "get *Biometrics* back on schedule and to recommend a new editor." The Committee asked Ralph A Bradley, an Associate Editor, to become the Acting Editor. Bradley and Treasurer Allyn Kimball went to Ottawa in mid-June to go over details with Hopkins; Bradley soon had the 1957 June issue with the printer and had hopes

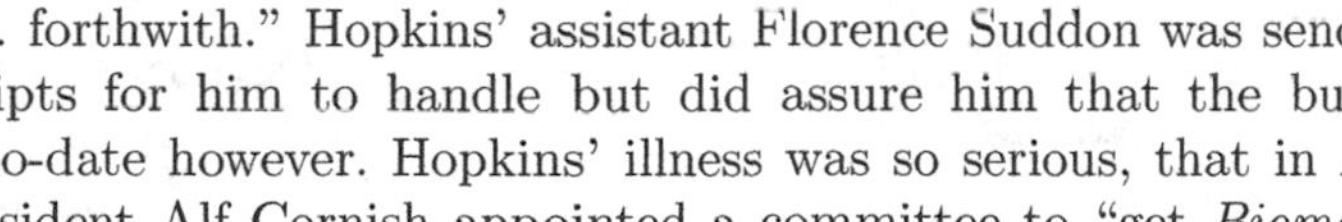

This editorial job must go to someone else real soon!

FIGURE 7.3
Part of Cox's letter to Bliss in November 1947.

to have the journal back on schedule early in the next year (actually achieved by the 1958 June issue). Almost immediately, Council approved Bradley as the next Editor.

Bradley in his 1961 Report, spelt out the disadvantages of having the Editor manage both the business aspects and the scientific aspects of being editor. He recommended that in future there be a Managing Editor and a technical editor. Therefore, upon completion of Bradley's term, Michael Sampford (BR) of Aberdeen Scotland was appointed for a five-year term from July 1, 1962 as the "Technical" Editor, i.e., the scientific editor. The business aspect of the journal would become the responsibility of a new appointment, the "Managing" Editor.

Each time, the appointment process for the next editor was always fraught with angst and not-inconsiderable activity. The President would appoint a Special Committee to make recommendations; "by tradition" Cox was always a member of this Committee. Lists of potential editors would be obtained. Then followed frank assessments of a particular person's suitability. Some would be seen as without tact, some "a little young" for the position, some as good at administration, some as not popular with (his) colleagues, one person once had a reputation as "quarrelsome" though he had reportedly "mellowed" these days, and so it went. Of course, these assessments also included some who were conscientious, and some who did have "tact," a seemingly *must* characteristic, who were subsequently considered further. Sometimes delays would occur; Figure 7.4 illuminates Finney's (May 10 and May 17, 1966) "response" to Cox when the search for a successor was too fraught with delays.

Technically, incoming editors were appointed by the President subject to approval of Council, following the advice of this Special Committee (appointed by the President). In later years, this process would involve the Editorial

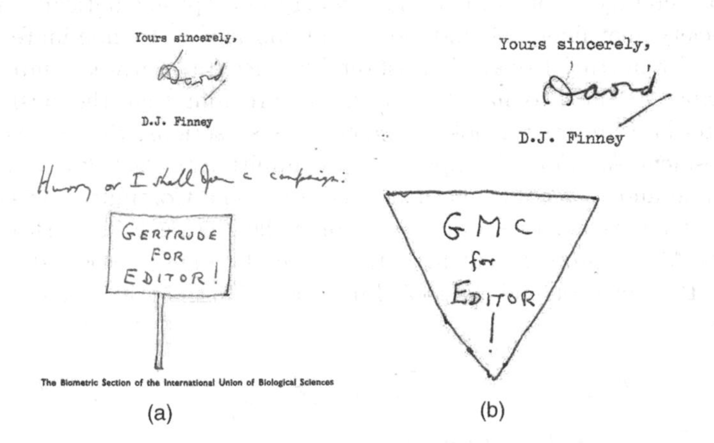

Yours sincerely,

David

D.J. Finney

Hurry or I shall open a campaign:

GERTRUDE FOR EDITOR!

The Biometric Section of the International Union of Biological Sciences

(a)

Yours sincerely,

David

D.J. Finney

GMC for EDITOR !

(b)

FIGURE 7.4
Search for the next editor – extract Finney's letter to Cox – (a) May 10, 1966 and (b) May 17, 1966 – Campaign Underway.

Board of which the Editor and Editors of other Society journals were ex-officio members. Shorter Communications Editors (and their earlier manifestation as Editor of Queries and Notes) and Book Review Editors were also appointed by the President subject to confirmation by Council. Associate Editors were appointed by the Editor.

Editorial policy statements would appear especially with the appointment of new editors. By and large though these tended to be consistent across editors. The first came from Cox in 1952 for the benefit of Editorial Board members. Then again later Hopkins felt the time had now come when such a policy should appear in the journal itself for the information of contributors and referees. Therefore, in consultation with the Editorial Board and the Society leadership, in 1956 we learn that "[*Biometrics*] general objectives are to promote and to extend the use of mathematical and statistical methods in pure and applied biological sciences, by describing and exemplifying developments in these methods and their application in a form readily assimilable by experimental scientists ... ."

Initially, Cox had an Editorial Board whereby regions selected their Board representative as an Assistant Editor (though the Editor retained the right to use, or not use, that person) with ENAR and WNAR appointing two each. The authors could submit their manuscripts to their region's Assistant Editor or to the Editor. This seemed to work well, though in handing over to Hopkins, Cox related an incident when an author submitted a manuscript to both and received opposite referee reactions.

By the time Frank Graybill became Editor in 1972, Editorial Board appointments and the various constraints had become confusing. The President could appoint twelve members, editors of journals published by regions were included, as well as Associate Editors. President Peter Armitage and Graybill wondered if any other Associate Editors appointed by Graybill could also be on the Editorial Board. Graybill saw the Editorial Board as a "watch dog and information center to help shape the journal on a continuing basis." This all led to Graybill's formal approach to Armitage suggesting it would be advantageous and timely to step back and look closely at the aims and structure of *Biometrics.* Accordingly, Armitage appointed a "*Biometrics* Committee" to survey the members, review editorial structure, review past and current editorial policies, and any other consideration deemed to be worthy of attention. This Committee consisted of Sam W Greenhouse (Chair, ENAR), L N Balaam (AR), Herb A David (ENAR, immediate past editor), and Hendrix de Jonge (ANed), with the 1974 President (C R Rao), Secretary (Hanspeter Thöni) and Editor (Frank Graybill) as ex-officio members. One upshot out of these deliberations was that the revised 1974 Constitution specified an Editorial Board consisting of at most twelve members appointed by the President for four-year terms. The Editor and Managing Editor and Editors of all journals published by the regions would be ex-offcio members. This Board was to advise the Editors, and to advise the President on successive editor appointments. All appointments here as elsewhere were to be approved by Council.

The 1983 constitutional revisions (see Chapter 10, Section 10.1) renamed the Editorial Board its Editorial Advisory Committee (EAC). Instead of twelve appointments from regions as in the previous Board, now all regions would have a member on the EAC whereby the members along with a Chair would be appointed by the President and subject to Council confirmation. The Nominating Committee (for new editors) of earlier years now became advisory to the EAC who in turn advised the President on appointments for the various editorial positions (Editor, Managing Editor where appropriate, Shorter Communications Editor and Book Review Editor) again subject to Council approval. These appointees, along with editors of regional journals, were also ex-officio members of the EAC.

Later, in 1997, an overall Executive Editor was appointed, the first (and only for a few years) being James A Calvin (1997—1999, ENAR), whose role was to oversee all editorial appointments. Raymond J Carroll (ENAR) was the new Editor and Louise Ryan (ENAR) was the new Shorter Communications Editor. Carroll also appointed an editorial assistant (Ann Hanhart, who was to so serve the then current and later editorial co-editors for twenty years eventually retiring in 2018).

In the 1990s, President Lynne Billard's charge to an editor selection committee, to make recommendations of successors to Editor Charles McGilchrist and Shorter Communications Editor Byron Morgan, had a contemporaneous twist. It was 1995 and the new computer capabilities and opportunities opening up from the superhighway revolution were descending upon the Society as an avalanche. It would be important that the next appointments were able to accommodate the demands of the new world as the move to a completely electronic submission process took hold (latex was the emerging consensus as the 'best' option but other electronic formats such as Word were deemed acceptable). Concurrently, however, it was important to be sensitive to those regions and authors who would not have the computer access and facilities of other regions. This covered editorial and production procedures as well as manuscript submission modes and was expected to result in a faster handling of papers. This was an issue that had begun to pervade many, ultimately all, components of the Society, not just publication outlets. Through the 1990s, computerization of all aspects of the journal process (from submission to the printed/electronic page on a member's desk) evolved with implementation occurring at different times as opportunities and changes of editorship occurred. Richard Tomassone as Editor of *Biometric Bulletin* in the mid-1990s led the way, quickly followed by *Biometrics* Editor Charles McGilchrist in January 1995.

The size of the journal was a frequent topic in the archives. While naturally editors wanted to accept any paper deemed publishable, the number of pages had obvious financial implications. Thus, editors necessarily had a balancing act to accept all that should be published but within the financial constraints;

requests for increased numbers of pages were not uncommon. Sampford was urged to stay with 900 pages so as not to exceed approved budgets. When Graybill was appointed, he was advised to work within a target of 1200 pages per year, raising standards if need be rather than "becoming narrower in subject matter." With the editor structural change in 1998, a policy of limiting papers to twenty-five pages or less was also implemented, as one avenue to reduce the backlog that had become a serious issue.

For the first few decades, the archives reveal regular reports from the Editor to the officers and to Council. A statement to the effect that the next issue would be on time, or not, would typically greet the reader. Often, because of varying circumstances usually beyond anyone's control, there would be delays; this always caused some level of consternation, as the timely mailing of the journal was a high priority. These reports typically ended with details of authors and titles of papers at various stages of the reviewing process – those accepted to appear, those rejected, and those still in the review pipeline.

7.5 Journal Management

At its inception, the management of *Biometrics* was carried out at the ASA office. The editorial management transferred to Raleigh effective from 1949, with Sarah Carroll doing the secretarial work as a type of (unofficial) Managing Editor. Then with the appointment of Hopkins as the new editor, the management was moved to Ottawa Canada as of January 1, 1956. Bradley became editor in mid-1967, and the business affairs were transferred to Bradley's base in Blacksburg Virginia by early 1958; and then again moved in 1960 to Tallahassee Florida after Bradley relocated there. Recall that since 1949 (see Section 7.2) all scientific and financial aspects of the journal were part of the duties of the Editor.

As indicated in Section 7.4 above, in his February 1961 Editor's Report to the Editorial Board, Bradley expanded on his belief that the time had come to have both a technical editor (as Editor-in-Chief, who handled the scientific aspects of the journal) and a Managing Editor (a Society member, who handled all journal business matters including sales of back issues and the handling of mailing lists for direct subscriptions), all duties at the time being carried out by the Editor. Bradley sent the same proposal to then-President Léopold Martin. It was unclear if this would entail changes in the By-laws, but in the end, it did not. Bradley proposed that such an appointment begin in the 1961 summer. This happened with the appointment in July 1961 of William A Glenn based at the Triangle Research Institute (Cox's domain) in North Carolina as Managing Editor. Bradley also said he would stay on for a year for this transition, but that he should then be replaced by a new editor. He concluded by opining that there should be term limits, e.g., five years, for future editors. These management proposals were adopted. Bradley also posed the notion that it was important that there be some financial reserves built up to ensure a safety net for the journal; while Bradley achieved a surplus of just over one year's operational costs, some advocated a goal of two years.

The previous year, Bradley's (April 1960) Report to the Editorial Board included details of securing a commercial firm to handle the maintenance of the mailing list for *Biometrics* to help reduce the burden on the Editor. After some stutters, Bradley working with Treasurer Allyn Kimball and then Marvin Kastenbaum engaged Mailing Services Incorporated in Richmond Virginia to maintain these lists. At the same time, since the stocks of back issues "had been shuffled about for years," this firm also became the repository for storing the back issues.

When Michael Sampford was approved as the Technical Editor, Malcolm E Turner of Richmond Virginia (soon moved to Emory University in Atlanta, Georgia, in 1963) was appointed as Managing Editor. Thus, this role was no longer traveling with the scientific Editor. The Managing Editor would be responsible for the business and financial side, including handling mailing lists, relations with printers, advertising, and the storage and sales of back issues. During this time, Turner arranged for the Copyright of *Biometrics*, starting with the December 1963 issue.

This seemed to work well, until 1968 when Turner was away on leave so that his secretarial assistant (Elaine Beckham) had to pick up the burden and constantly had to involve the then Editor Herbert A David. By November 1968, President Cox declared the "situation is impossible." As it so happened, discussions had began earlier in the decade to establish a more permanent business office with a Business Manager whose duties would include those of the journal's Managing Editor, along with duties on the business side of the Secretary's and Treasurer's office. In mid-1968, Cox gave a heads-up to Turner though Cox expected the change would not occur until at least 1969. In late November 1968, Cox explained to Turner that this new office would indeed happen. Then, quite suddenly, in December 1968, Turner resigned (and also from the Society) as did his assistant to take up a new publishing venture of their own. Although Editor David and President Cox had praised their work, Turner lashed out seemingly taking offence. In the end however, Beckham continued to help with the transition for several months into 1969. Thus, it was that in the early months of 1969, a Business Office was formally established, in Raleigh, to take over the non-scientific duties of the journal. Larry Nelson was the first such Business Manager; Nelson later served as the Society Treasurer (1970—1978). With this move, the title "Technical Editor" once again became the "Editor." In 1970, mailing of *Biometrics* was facilitated by a "single computerized list."

As an aside, this was not the first, nor last, time that an occupant of some position did not welcome such changes. Over the years, there were occasions when occupants (sometimes a member, sometimes a hiree) developed a propriety sense of some position. Indeed, Bliss when stepping in to make adjustments where needed, would express his firmly held belief that regardless of office, term limits were important – not just to overcome an inherent jaded-ness that came with too-long an occupancy, but more because he strongly felt that it was important to share duties among as many members as possible to keep a vibrancy in the level and growth of the Society. On this occasion, it was the Managing Editor, though interestingly this was somewhat an anomaly since an unspoken (sometimes spoken) sense of relief was expressed by previous editors when their term was up, relief primarily because of the burden of these non-scientific aspects of the position.

Editor Daniel Solomon in the 1980s added a Production Editor (Susan Reiland) to assist him in preparing the manuscripts for the publisher, continued by the next two editors. However, in 1993, Reiland requested effectively a 50% increase in compensation, a considerable amount. Production issues overall had become a problem. Therefore, in December 1993, President Niels Keiding appointed a Production Committee (Chair, Vice-President Billard, ENAR) to study all aspects of journal publication broadly. An interim 1994 report sought more time so that relevant advances related to computer technology could be more extensively considered. Further, this Committee wanted to work in conjunction along with newly appointed (by Billard) ad hoc Electronic Media Committee (Chair, Éric LeBoulengé, RBe) and ad hoc Technology Committee (Chair, Adrian Bowman, BR), as they wrestled with the largely unknown forces necessary to bring the Society's publications into this new electronic future. Though recognizing the fundamental truism that uniformity of submission should not be enforced, now-Editor Charles McGilchrist wanted to have electronic files for papers because, among other issues, he felt this would put the journal ahead in time (and simultaneously thus alleviate the backlog issue) and would assure better accuracy of page proof preparation. Meantime, though Reiland's work was deemed to be excellent, she felt under-appreciated and resigned at the end of 1994 (shades of the Technical Editor in the 1960s); this allowed McGilchrist to change immediately (in January 1995) to producing page proofs from disks. The Shorter Communications Editor Byron Morgan had already shifted to electronic modes. When the editorial structure was changed in 1999 to a three set of rotating editors (see Chapter 11, Section 11.2), the move was made to use the internet and electronic filing procedures to optimize submission and production procedures.

Meantime, concurrently, Society business operations were transitioning (in 1994) to a professional organization management firm whose functions subsumed the roles of the Business Manager including the business aspects of journal publications; see Chapter 10, Section 10.2). This new management quickly ascertained that the Society could benefit from a reassessment of its production of *Biometrics*. These observations dove-tailed with the deliberations of the Production Committee; and also the Electronic Media Committee and the Technology Committee though the purview of these latter two committees went beyond the journal alone. Thus, with a focus on

Biometrics, eventually, during 1995, proposals were "solicited from interested companies to provide detailed information as to editorial, production, printing and mailing ... and at what cost." The final report in 1996 provides more details of the committee's deliberations, concluding with the news that the selected company with the March 1996 issue of *Biometrics* generated over $11,000 in savings (compared to a projected $7000—$10,000 per issue). The production role was overseen by the new International Business Office, with the scientific editorial roles maintained within the Editor's domain. [This also applied to the *Biometric Bulletin* (discussed in Chapter 8, Section 8.1) thus making it easier to recruit successor editors.]

The sales of back issues constituted important sources of income in early decades. It is recalled that how these were to be handled was even a component of the Agreement to transfer the journal from the ASA to the Society; see Section 7.2. Indeed, additional printings of old issues sometimes occurred to meet anticipated needs. However, by the 1980s, costs of the storage of back issues exceeded the financial benefits of sales. The Awards Fund Committee (see Chapter 3) became an active player, and regions, especially ENAR, studied the issue and soon adopted a policy of covering the costs of sending back issues to libraries in developing countries, to help unload what had become a financial liability. Sometimes grants supported the cost of printing conference proceedings in the journal, e.g., financial assistance from the International Union of Biological sciences (IUBS) helped defray the costs of proceedings from the Campinas Brazil Symposium of 1955, as had Rockefeller Foundation monies helped defray the costs of the Woods Hole proceedings.

As expected, there were many archival materials relating to production issues. Some dealt with printing; these discussions revolved around who should print the journal, be this an outfit either in North America as for early volumes, but the possibility to have non-American printers was also explored as needed. It is instructive to note that the chosen printer did indeed change over time. Another issue that received much attention was mailing. Postal service regulations differed in their requirements and attributes across countries, particularly for undeliverable mail (members did move constantly!). Postal workers' strikes, longshoremen's strikes among other complications, necessitated temporary alternative arrangements. The question of when and where and by whom would come up on a regular basis. Estimates and samples from prospective printers would be obtained and decisions made. During Cox's term, sales of reprints balanced reasonably well with printing charges.

Concerns, even consternation, about back-logs surfaced at various times. This first appeared in 1965, with discussions about increasing the number of pages to handle this problem. However, increasing the size of an issue had considerable financial implications and so was strongly discouraged. The situation had become so bad in the early 1990s that ENAR had considered helping out. With a two-year backlog in 1994, and recognizing that about 70% of recent author-ships were based in ENAR-WNAR, one suggestion was to add a surcharge to those regions's dues to reduce this "disaster" backlog.

To add additional pages and/or an extra issue would have obvious financial implications (subjected too by large inflationary forces), as well as impacting arrangements with publishers and postal services and the like. There would be suggested solutions proposed, debated, and ultimately accepted or dismissed. The year 1993 saw some thoughts about starting a *Biometrics B* focusing on medical papers as a way to alleviate the backlog problem; but this idea did not gain traction. A related issue was the time a manuscript spent in the reviewing pipeline. Moving to electronic procedures helped reduce these times.

7.6 Importance

As intimated earlier, *Biometrics* was the scientific backbone of the Society. Many members joined precisely because of this journal. Dues structures were fashioned around the costs of its production. Deliberations surrounding the issues of regional and national group formations, particularly in developing world countries, as often as not, were concerned with how the journal could be made available at a cost that balanced actual costs of producing the journal and local currency restrictions and the like; see Chapter 6. That said, not surprisingly then, the Archives contained an inordinate number of letters and communications between the general leadership (mostly Secretary and/or Treasurer) and regional and national group officers detailing that "so-and-so" was in arrears. The paid-up cut-off date varied occasionally; but typically dues paid before the first issue of the journal was mailed out meant that the member was classified as a current member, while for those who paid later it meant that membership became effective from the next January. In his report to Council in November 1957, Secretary Healy related how some members would miss one year's dues and subsequently resume payment, "an unsatisfactory state of affairs ... should not be allowed except under exceptional circumstances (and that only authorized by a Regional Committee)." The British Region even revised their Regional By-laws to address this question, in effect allowing for a maximum two-year abeyance during which they would not receive *Biometrics* nor enjoy other benefits of membership. Outside of these special circumstances, any subsequent payment of dues was "regarded as paying off these arrears."

8

Later Publications

While *Biometrics* was, is, the scientific backbone of the Society, other publication proposals were also considered and adopted over the years. In 1984, a new newsletter *Biometric Bulletin* was launched; see Section 8.1. At various stages, complaints that *Biometrics* had become too mathematical variously defined (see Chapter 7, Section 7.3) led eventually to the publication of the *Journal of Agricultural Biological and Environmental Statistics (JABES)*, a more application-driven outlet of scientific endeavors; this is described in Section 8.2. Then, in Section 8.3, we cover other scientific publications (Section 8.3.1), regional publications (Section 8.3.2), and non-scientific publications (Section 8.3.3). Finally, for the record, in Chapter 12, Section 12.3 provides a list of those who so served the Society as Editors of these different journals.

8.1 *Biometric Bulletin*

During President Herb David's tenure, the idea to introduce a new publication, which subsequently became the *Biometric Bulletin*, emerged. The Editorial Advisory Committee (EAC, the newly named former Editorial Board under the new Constitution, chaired by Robert Curnow, BR) was asked to develop plans and proposals to be sent to the Executive Committee and hence to the Council for input and discussion. It was determined that the *Biometric Bulletin* would contain Society news and announcements including those for regions, member news, non-technical letters to the Editor (scientific letters to the Editor would stay with *Biometrics*), and some advertisements (as for *Biometrics* still). In other words, the newsletter component of *Biometrics* was to be moved to the *Biometric Bulletin*.

After a certain amount of equivocation from various folk and "complicated!" discussion, Peter Armitage (1979–1984 Editor of *Biometrics*) settled the issue of abstracts of conference papers expressing clearly a strong preference that these also be transferred to the *Biometric Bulletin*; Council at the 1984 International Biometric Conference (IBC) suggested the abstracts be included if their costs were at or below the costs of publishing them in *Biometrics* (achieved finally in 1987). Thus, the editor had a 150-word limit and

DOI: 10.1201/9781003285366-8

a stipulation that the abstract was not published elsewhere; by 1994, the word limit was 50. There had been some discussion about moving the Book Review section as well, but ultimately it was decided to keep this in *Biometrics.*

This was a quarterly publication; the first issue (see Figure 8.1a) appeared in May 1984 with Robert O Kuehl (WNAR) as the first editor. The editor would be an ex-officio member of the EAC. Within four years, the *Biometric Bulletin* was seen as "an invaluable vehicle" for timely dissemination of information of the Society and its Regions and National Groups.

Since ballots were inserted into issues, an early complication arose when it was realized that student members would receive a ballot though they were not eligible to vote, and that associate members (who were eligible to vote) did not receive the *Biometric Bulletin* and so did not have the ballot.

Initially, Editorial Board members were to serve as their respective Regional Correspondents to this new publication outlet; national groups were also asked to provide a correspondent to represent their group members. Later, these correspondents were appointed by the respective regions or national groups.

As for *Biometrics*, editors were responsible for finding suitable publishers and the like. Thus, there were lots of letters between the Editor and the Society about publisher choices, associated costs of printing (and by which printers; the film negatives created by the printer were by trade practice the property of the printer, and destroyed after two years), costs of mailing (surface or airmail), the cost and time to reach members. At times, issues arose with the suitability or not of address labels. On one occasion, the operative printing company became insolvent necessitating obvious changes. Equipment to perform their editorial duties was often 'donated' by the editor's institution to varying degrees. However, when "the printer at Rothamsted ha[d] resigned," Editor Joe Perry (BR) did not think printing with letterhead could be done at Rothamsted.

Perry was thoughtful. For example, he raised the question of including advertising to help defray costs; but, if so, then there should be a limit on how much advertising could be included as well as a disclaimer to the effect that "the views were not those of the Society or the publisher" should be added; plus Perry suggested that taking out indemnity insurance against defamation should also be considered. Having determined that its inclusion would no longer affect the second-class postal permit, advertising started in 1988. This was a successful addition. Color was also added in 1988 to help attract advertising (see Figure 8.1b). However, after a few years, although it had became a useful income-producing portion, balance between overly too much advertising became an issue at times while still maintaining the primary purpose of the *Biometric Bulletin* to inform members of Society activities. Perry also wondered if obtaining a copyright should be considered, as well as a logo. These came to be in the mid-nineties with the move of the Society management to a professional association management firm (see Chapter 10, Section 10.2).

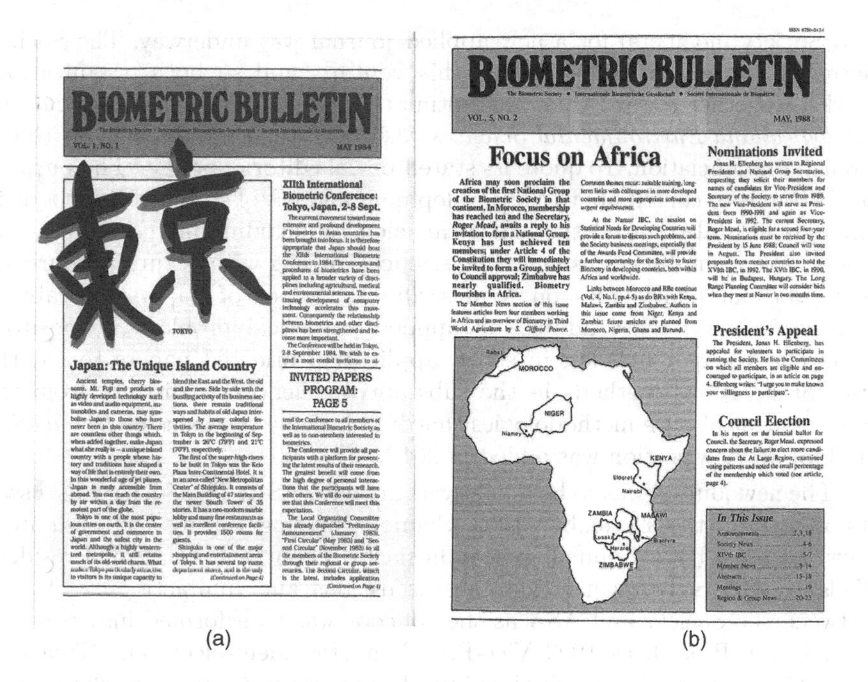

BIOMETRIC BULLETIN

VOL. 1, NO. 1 MAY 1984

東京

TOKYO

Japan: The Unique Island Country

Ancient temples, cherry blossoms, Mt. Fuji and products of highly developed technology such as video and audio equipment, automobiles and cameras, may symbolize Japan to those who have never been in this country. There are countless other things which, when added together, make Japan what she really is – a unique island country with a people whose history and traditions have shaped a way of life that is entirely their own. In this wonderful age of jet planes, Japan is easily accessible from abroad. You can reach the country by air within a day from the remotest part of the globe.

Tokyo is one of the most populous cities on earth. It is the center of government and commerce in Japan and the safest city in the world. Although a highly westernized metropolis, it still retains much of its old-world charm. What makes Tokyo particularly attractive to visitors is its unique capacity to blend the East and the West, the old and the new. Side by side with the bustling activity of its business sections, there remain traditional ways and habits of old Japan interspersed by many colorful festivities. The average temperature in Tokyo in the beginning of September is 26°C (79°F) and 21°C (70°F), respectively.

The first of the super-high-risers to be built in Tokyo was the Keio Plaza Inter-Continental Hotel. It is in an area called "New Metropolitan Center" of Shinjuku. It consists of the Main Building of 47 stories and the newer South Tower of 35 stories. It has a neo-modern marble lobby and many fine restaurants as well as excellent conference facilities. It provides 1500 rooms for guests.

Shinjuku is one of the major shopping and entertainment areas of Tokyo. It has several top name department stores, and is the only

(Continued on Page 4)

XIIth International Biometric Conference: Tokyo, Japan, 2-8 Sept.

The current movement toward more extensive and profound development of biometrics in Asian countries has been brought into focus. It is therefore appropriate that Japan should host the XIIth International Biometric Conference in 1984. The concepts and procedures of biometrics have been applied to a broader variety of disciplines including agricultural, medical and environmental sciences. The continuing development of computer technology accelerates this movement. Consequently the relationship between biometrics and other disciplines has been strengthened and become more important.

The Conference will be held in Tokyo, 2-8 September 1984. We wish to extend a most cordial invitation to attend the Conference to all members of the International Biometric Society as well as to non-members interested in biometrics.

INVITED PAPERS PROGRAM: PAGE 5

The Conference will provide all participants with a platform for presenting the latest results of their research. The greatest benefit will come from the high degree of personal interactions that the participants will have with others. We will do our utmost to see that this Conference will meet this expectation.

The Local Organizing Committee has already dispatched "Preliminary Announcement" (January 1983), "First Circular" (May 1983) and "Second Circular" (November 1983) to all the members of the Biometric Society through their regional or group secretaries. The Second Circular, which is the latest, includes application

(Continued on Page 4)

(a)

BIOMETRIC BULLETIN

The Biometric Society • Internationale Biometrische Gesellschaft • Société Internationale de Biométrie

VOL. 5, NO. 2 MAY, 1988

Focus on Africa

Africa may soon proclaim the creation of the first National Group of the Biometric Society in that continent. In Morocco, membership has reached ten and the Secretary, *Roger Mead*, awaits a reply to his invitation to form a National Group. Kenya has just achieved ten members; under Article 4 of the Constitution they will immediately be invited to form a Group, subject to Council approval; Zimbabwe has nearly qualified. Biometry flourishes in Africa.

The Member News section of this issue features articles from four members working in Africa and an overview of Biometry in Third World Agriculture by *S. Clifford Pearce*. Common themes recur: suitable training, long-term links with colleagues in more developed countries and more appropriate software are *urgent requirements*.

At the Namur IBC, the session on Statistical Needs for Developing Countries will provide a forum to discuss such problems, and the Society business meetings, especially that of the Awards Fund Committee, will provide a further opportunity for the Society to foster Biometry in developing countries, both within Africa and worldwide.

Links between Morocco and RBe continue (Vol. 4, No.1, pp.4-5) as do BR's with Kenya, Malawi, Zambia and Zimbabwe. Authors in this issue come from Niger, Kenya and Zambia; future articles are planned from Morocco, Nigeria, Ghana and Burundi.

Nominations Invited

Jonas H. Ellenberg has written to Regional Presidents and National Group Secretaries, requesting they solicit their members for names of candidates for Vice-President and Secretary of the Society, to serve from 1989. The new Vice-President will serve as President from 1990-1991 and again as Vice-President in 1992. The current Secretary, Roger Mead, is eligible for a second four-year term. Nominations must be received by the President by 15 June 1988; Council will vote in August. The President also invited proposals from member countries to hold the XVIth IBC, in 1992. The XVth IBC, in 1990, will be in Budapest, Hungary. The Long Range Planning Committee will consider bids when they meet at Namur in two months time.

President's Appeal

The President, Jonas H. Ellenberg, has appealed for volunteers to participate in running the Society. He lists the Committees on which all members are eligible and encouraged to participate, in an article on page 4. Ellenberg writes: "I urge you to make known your willingness to participate".

Council Election

In his report on the biennial ballot for Council, the Secretary, Roger Mead, expressed concern about the failure to elect more candidates from the At Large Region, examined voting patterns and noted the small percentage of the membership which voted (see article, page 4).

In This Issue

(b)

FIGURE 8.1
Biometric Bulletin: (a) first issue May 1984, and (b) color added May 1988.

During the tenure of Richard Tomassone (RF) as Editor (May 1994 to December 1997), electronic communications were taking hold, although earlier in 1987, Kuehl reported that input to typesetting machines had been completely converted to microcomputer diskettes. In particular, while sensitive to difficulties for some members/regions, Tomassone requested materials be sent by email, to be transferred to diskettes and hence formatted for production.

8.2 *Journal of Agricultural Biological and Environmental Statistics (JABES)*

By 1990, perennial complaints about mathematical sophistication in *Biometrics* expanded to include criticisms of the narrow focus on medically and epidemiologically oriented applications and especially clinical trials research. Thence, in 1990, an ad hoc committee to look at this issue was formed, members were polled, and by 1992, it was clear that support existed for a second more applied journal. Concurrently, an American Statistical Association (ASA) initiative led by Brian Marx and Linda Young (both ASA and Society members) for a new applied journal was underway. The Society accepted the invitation to join in this venture, and so both organizations worked together toward what was ultimately named *Journal of Agricultural Biological and Environmental Sciences (JABES)*; the acronym was set to the French pronunciation. To quote its stated initial editorial policy "The purpose of *JABES* is to contribute to the development and use of statistical methods in the agricultural sciences, the biological sciences including biotechnology, and the environmental sciences including sciences dealing with natural resources." The editorial policy goes on to describe the types of papers desirable – interdisciplinary, those addressing applied statistical problems, expository review and survey articles; in short, application oriented papers to aid the user of statistical methods in the substantive sciences as distinct from the development of pure methodologies that had become the focus of *Biometrics.* A Book Review section was added in 2012.

The new journal was to be a joint venture with ASA. The Society discussed its merits at the December 1992 Council meeting during the Hamilton, New Zealand IBC. Council was enthusiastic about the venture. President Niels Keiding subsequently led the discussions and interactions with and between the Society and ASA as the concept was transformed into concrete fact; Lynne Billard, as 1993 Vice-President (President-elect) and Treasurer Steve George were also heavily involved. In particular, George was the Society point man in developing the financial aspects of the initial five-year plan. The Society worked steadily toward an agreement that was equitable throughout. For example, the ASA had suggested at first that they provide the start-up funds, but as the Society wanted to be truly equal partners ultimately both organizations contributed equally. Equity also included journal management in that both societies appointed two members to the Management Committee with a fifth person belonging to both societies to serve as Chair; this is different in structure from ASA's journal management committees.

Thus, it was that both organizations would share equally in financial contributions and not just in initial start-up funds and the like. However, once underway, once established with start-up funds repaid, the equitable distribution of profits and/or losses would be proportionate to the numbers of subscribers from the respective organizations. The production itself was handled by the ASA office. A clause, reminiscent of the transfer of *Biometrics* from the ASA to the Society (see Chapter 7, Section 7.2), allowed for either party to opt out (with a year's notice) after five years.

The Management Committee selected the Editor who would in turn be approved through the processes of both organizations (for the Society, this was a President's appointment based on the advice of the Editorial Advisory Committee and subject to the approval of Council). This Committee would make recommendations regarding pricing, marketing, and related issues, including consideration of pricing for members in transition and developing countries. Committee members would serve three-year terms. The first Management Committee was chaired by Linda Young (who later became

2007–2013 Treasurer, ENAR). The first Editor was Dallas E Johnson, a member of both the Society (ENAR) and ASA. Not necessarily a requirement, but all 21 associate editors were also members of the Society from nine Society countries.

The early goal was to have the first edition ready for 1994. In the end, this was deferred to 1996, though Johnson began taking submissions in 1994. In an odd twist, by the time the Agreement was settled to both ASA and the Society's satisfaction, Billard was a signatory to the joint venture agreement, not as a Society officer but as the then-ASA President[!] along with the then-IBS President Byron Morgan. Figure 8.2 shows this Agreement.

While clearly a quality journal, the first few years saw losses beyond those envisioned in the planning stages. How these were to be handled had been clearly spelled out in the discussions between the two organizations and in the 1996 Agreement. Unfortunately, during this time, there was a large turnover of key ASA personnel originally involved with the financial details. In particular, the new ASA contact person (at the time, relatively inexperienced, and unfamiliar with the earlier negotiating aspects) was interpreting agreement terms at variance with the original intent and seemed unable to produce accurate and timely information further complicating matters. The 1998—1999 Society Archives detail extensive Management Committee records, showing a lot of back and forth and financial calculations (primarily between Jonas Ellenberg, Billard, Young as Management Committee Chair, and President Susan Wilson but also Treasurer George) clearly demonstrating the need for a revised agreement between the two organizations. Ultimately in May 1999, a Revised Agreement was signed by ASA President Ellenberg (who in prior years as a Society Treasurer and Society President had been involved in earlier discussions of an applied journal) and our Society President Wilson. The overriding general principles remained intact, though each organization increased their original dollar investments to cover initial costs, and responsibility for the management was moved from ASA to the Society. The original Agreement was but two pages; the new Agreement was thirteen pages, such were the determined efforts to remove any ambiguities of interpretation!

8.3 Other Publications

8.3.1 Scientific Publications

With Fisher's death, colleagues from Adelaide[1], most especially J Henry Bennett, started gathering Fisher's papers and materials for archival purposes,

[1] Upon his retirement in England, Fisher had re-located to Adelaide Australia to work with Cornish at the Commonwealth Scientific Industry and Research Organization (CSIRO). It was here that he died on July 29, 1962. See Billard (2014); extracts from that article's Supplementary Materials with details of letters written by Cornish and Bennett on Fisher's death are provided in the Appendix of this Volume.

Agreement Between the International Biometric Society and the American Statistical Association
Concerning the Journal of Agricultural, Biological, and Environmental Statistics

The American Statistical Association (ASA) and the International Biometric Society (IBS) agree to jointly sponsor the *Journal of Agricultural, Biological and Environmental Statistics* (JABES). A joint venture will be set up so that the funds for this journal will be segregated from the funds of either sponsoring society. There will be a professionally audited yearly financial report sent to both societies. There will be joint ownership.

There will be a JABES Management Committee (MC) with two members appointed by IBS, two members appointed by ASA, and a fifth member, appointed jointly by ASA and IBS, to chair the JABES MC. The JABES MC will serve as a Search Committee for new Editors at the appropriate time and will set editorial policy. After the original start-up, the members of the JABES MC will serve three-year terms with appointment for one additional three-year term possible. The initial JABES MC members will have staggered terms so that a complete turnover of members will be avoided. All Editor selections are to be ratified by the governing bodies of each society.

The JABES MC will review and recommend subscription prices, reprint fees, page charges, complementary copies, back-issue sales, and other such issues. These recommendations will be ratified by the governing boards of both societies each year by September 1. IBS, through its representatives on the JABES MC, will make recommendations on the price for the journals in countries in transition and developing countries. The chair of the JABES MC will sit on the ASA Journals Management Committee and the IBS Editorial Advisory Committee.

ASA and IBS will advance start-up funds for the journal equally. Surpluses and deficits will be divided in the same proportion as gross revenue from subscribers. For every subscriber who is a member of both societies, subscriber revenue will be divided evenly. For subscribers who belong to only one society, revenue will be attributed to that society. The percentage of overlapping members will be calculated by a method approved by the JABES MC. Library subscribers' revenue will be divided in the same proportion as member subscriber revenue.

ASA will manage the production of the journal. A five-year production, marketing and budget plan will be developed jointly for the journal. This plan will detail the number of issues, production costs, marketing plan and costs, editorial costs, potential number of member subscribers, and costs of maintaining subscriptions and filling orders. ASA will handle the audit, subscription records, accounting records and the quarterly statements. This plan will be sent to the JABES MC for comment and approval and then to the governing boards of each society for approval and information. These estimates will be used to set the initial price of the journal for member subscribers, individual nonmembers, and libraries. There will be no advertising in the journal.

Cash distributions will be made in proportion to accrued surpluses to each society only after the equivalent of 100 percent of the next year's operating expenses have been reserved.

After the first five years of full production, either society may discontinue its participation upon a one year notice to do so. The remaining partner may buy out the other society at a value to be determined by the most recent audit and for the member percentage of equity based on the most recent subscription rates from the two societies.

KBillard — 6/21/96
(President - ASA) (date)

B.J.D. [signature] — 1/7/96
(President - IBS) (date)

FIGURE 8.2
JABES Agreement between the society and ASA.

all his scientific works but not reviews. In particular, these would take the form of a five-volume set edited by Bennett and former Society and Australasian President Alf Cornish. The editors wanted 10% of the royalties to go to the Fisher Memorial Fund in Adelaide. Naturally, the Society was approached, in 1965, to see what interest, if any, and to what degree, it might have. Oliver and Boyd would be the publisher. The Society was interested, and Council approved. The Society was not worried about any precedent being set as it would be some time before there would be someone else "of the stature of R. A. Fisher to honor so." Thence, communications traveled between the Society and Bennett to ascertain details. Would the Society provide a loan (or even a contribution, and to what amount) to allow this to proceed, what were the legal implications, who would share in the profits, the questions were many. The proposal to Council was that the Society help underwrite the effort by paying the publishers US$2000 to be paid when the galleys arrived and to be reimbursed (as sales progressed). Then-President David Finney was in no doubt that this was a desirable goal but he did wonder if the Society's small reserves could absorb this outlay in the perceived unlikely event of losses.

As an entirely different venture, Oxford University Press approached President Geoffrey Freeman in 1987 to explore the possibility of a series of what came to be named as a Biometric Monograph Series. Any proposal would have to be approved by the publisher; and Monograph Editorial Board members should be senior members of the Society. The Society began discussing the merits of this venture. Incoming President Ellenberg asked Freeman to liaison with the publisher and to work with the Editorial Advisory Committee. They were to bring a proposal to the Finance Committee and Council at the Namur 1988 IBC. Given the favorable responses, subsequently, an ad hoc committee appointed in 1992 by President Keiding prepared a Final Report in 1993. The 1994 Revised Final Report received Council approval and Nicholas A Lange (ENAR) was appointed as the founding editor. Details still had to determined. The aim of the Series was to provide a forum devoted to a single topic most especially in non-medical areas, though it could be written by one or a few authors. Analogies and contrasts with the successful ENAR-WNAR initiative *Case Studies in Biometry* (CSB, see Lange et al., 1994) were drawn. As for CSB whose royalties went to ENAR and WNAR, royalties for the monographs would go to the Society. Ultimately, although a founding editor and successor editors were appointed, as well as an editorial advisory board, the Series itself never materialized; it was not for want of effort however.

In celebration of the first fifty years of the Society (forty-nine years for the Society and fifty-one years for *Biometrics*), a volume *Advances in Biometry* was published, edited by Peter Armitage and Herbert A David, both former *Biometrics* editors and both former Society Presidents; see Armitage and David (1996). Articles were solicited, reviewed, and then accepted or otherwise for publication.

The Society debated at length whether to be part of a journal *Biostatistica* begun in 1990, designed to abstract articles based on applications of statistics to biology from a wide range of journals. This would include abstracts from all IBCs and Regional meetings. In the end, this proposal was dropped.

8.3.2 Regional Publications

In 1953, the Belgian Region (RBe) had proposed a second journal *Acta Biometrica* to the Society in response to the perceived too-mathematical slant dominating *Biometrics* articles. Though debated at length, agreement on certain details was too elusive; see Chapter 7, Section 7.3. Subsequently, the Region started the journal *Biométrie Praximétrie* in 1960 as a regional undertaking. The first editor was Léopold Martin (who had proposed the journal, and who also served as Society President in 1960–1961). It was multilingual (French, Dutch, English, and German), published quarterly, until publication ceased in 1994 due to financial problems.

The German Region published the journal *Biometrische Zeitschrift*, co-founded in 1960 by Ottokar Heinisch and Maria-Pia Geppert, as co-editors-in chief. Upon Heinisch's death in 1966, Geppert and Erna Weber became co-editors-in chief, after which Erna Weber assumed this role. It is published six times a year. As for *Biométrie Praximétrie*, papers focus on application-driven innovations, in this case primarily from medically and environmentally related fields. The journal's title progressed first to *Biometrische Zeitschrift – Biometrical Journal* and then to today's *Biometrical Journal*.

8.3.3 Non-scientific Publications - Directory

A plan to publish a list of members in the December 1948 issue of *Biometrics* was shelved to publish instead a Year Book in 1949, containing lists of members and officers, the Constitution, Council, Regional By-laws and a calendar of activities, and to be available to members only. This became the first Directory which appeared in 1949 and included "850 or more members and a geographical index." This was free to members but was available at cost to non-members. In those days, at the time of joining the Society, applicants identified their sub-field of scientific interest; this information was also included, and details of their distribution presented. For example, in 1949, 27.4%, 32.9%, 16.2%, 14.1%, and 9.3% were in the mathematics and statistics, general and applied biology, medical sciences, human biology and public health, and other categories, respectively. The second Directory appeared in 1953 reporting on 1144 members (including "about 100 women" interested primarily in teaching and research); this directory was prepared with IBM punch cards under the direction of Colin White (Assistant Secretary-Treasurer while Chester Bliss was on leave). Records showed the growth of membership across the years. However, this 1953 Directory stated that there had been a net 25% increase since the 1949 Directory, even though 40% of those from 1949 were no longer members.

In general, the Directory would contain lists of current regional, national group and Society officers as well as past Society Presidents, and the current Society Constitution and By-laws. Importantly, it listed names and addresses of Society members including their region. Prior to its publication, there would be lots of communications between the Secretary and the regional and national group secretaries to insure accuracy and up-to-datedness. Members often provided their own details in hand-written communiques which were not always readable, among other problems that occupied the attention of officers involved. For those who have heard Finney's oft-repeated mantra on this, it is no surprise to read about his preoccupation with peculiarities of different languages such as diacritical marks, even as early as 1964, though Finney did concede "we shan't achieve perfection." Nor, as he reminded the participants at the fiftieth Anniversary Celebration at the 1996 IBC with the Directory in hand (see Figure 8.3), was it achieved in the 1996 Directory.

These were published approximately every three years. As to be expected, the archives contained considerable communications between the Treasurer (initially, later the Business Office, although for 1980 it was both the Treasurer and the Business Office) and Regional Secretaries and National Secretaries as lists of members were to be submitted in a requested format. Inevitably, delays would occur when one region or another was slow in providing the necessary

FIGURE 8.3
1996 directory – front cover.

information and/or had not returned the corrected proofs in a timely fashion. Once published, there would be letters from individuals or their regional officers with complaints about inaccuracies relating to entries. The omission of a Sustaining Member was always a cause of consternation; a followup letter of apology would be sent. Nevertheless, publication of the Directory was a massive undertaking, eagerly awaited, and well received once available.

Other letters would revolve around questions of when the Directory would be ready. Still other materials were concerned with the cost of producing the Directory, and where it might be printed. This also covered debates about the charge to be made to non-members who wanted copies. Indeed, while members saw the Directory as a very useful source to keep in touch with each other, outside entities found it useful for a variety of reasons including being sold as mailing lists to interested parties. However, after 1972, the policy of selling mailing lists was discontinued.

9

Scientific Meetings

Close on the heels of *Biometrics* was the importance of meetings, symposia and conferences, as the prime medium for discussion and dissemination of scientific ideas. Indeed, the Society itself was founded at one such meeting, the First International Biometric Conference (IBC). Since then, IBCs have occurred regularly; these are covered in Section 9.1. Further, in addition to the IBCs, there have been many important meetings, key in the growth of the Society, especially in the early decades. Unarguably, the most successful of these was the Brazil Symposium held in 1955 in Campinas, an IBC in all but name, but there were others; see Section 9.2. These sections focus on the archival records of the Society's first fifty years; a brief summary of international meetings after 1997 is described in Chapter 11, Section 11.3. Where possible, the regional and/or national group affiliations of key participants have been included; these abbreviations are summarized in Chapter 12, Table 12.1. A summary of locations and organizing chairs and program chairs where applicable for these IBCs is given in Chapter 12, Section 12.4, Table 12.2.

The archival records as they deal with conferences, and IBCs in particular, overflow with what might be called trivia as details pertaining to accommodation costs, transportation questions and costs, and other practical aspects of holding a meeting are conveyed. Separate communications dealt with potential speakers – for the scientific program; not to be forgotten was an accompanying social program. Initially, these are included in back-and-forth letters (no email, phones were expensive and so sparingly used, no FAX, etc.) between Society officers. Eventually, dates and locations, along with relevant details, were secured. Thence, notices would be sent to the membership through announcements in the journal *Biometrics* as well as additional communications to members (often through the regional leaderships). With its launch in 1984, the notices moved to the *Biometric Bulletin*. Some, but by no means all – indeed only a mere smattering – of these details will be reported here, mainly to convey a sense of the similarities of the core scientific intentions but also of differences, especially those between costs, in the early years compared to those of today. An attempt is made herein to convey the flavor of these records rather than simply the sanitized published versions.

In those early decades, an important driver for a location of an upcoming IBC was the timing and location of an International Statistical Institute (ISI) Session held biennially. While this was an absolute necessity for the first IBC (see Section 9.1.1), it was also a key factor for subsequent IBCs. Therefore,

DOI: 10.1201/9781003285366-9

the archives contained a lot of material pertaining to the ISI and its future sessions. Further, since many members worked in genetics[1] or related areas, the times and locations of major genetics conferences also played a role in setting up future IBCs.

9.1 International Biometric Conferences

9.1.1 IBC 1 – Woods Hole USA, 1947

The First IBC at Woods Hole was held (September 5–6, 1947) immediately preceding the twenty-fifty ISI Session slated for September 6–18, 1947 in Washington, DC.

Technically, this was an IBC, a scientific conference, with Chester Bliss as Organizing Chair. However, the story of this IBC is inextricably intertwined with the story of the formation of the Society. An organizational session was held at the opening session on the first day, Friday September 5, at which proposals for a new society were outlined and relevant committees formed. The Saturday morning session innocuously titled "International Cooperation in Biometrics" was the key session during which the Society was born and a constitution adopted. As seen from page 2 of the Scientific Program displayed in Figure 9.1, scientific sessions occurred Friday afternoon and Saturday afternoon. The complete four-page Conference Flyer announcing the IBC is provided in Figure 9.1. Thus, we learn that registration was free but required, accommodation was to be in dormitories at $1.50 per person per night (though other arrangements were possible "in the village"), meals were $2.00 per person per day; and the social event was a clam bake on the Friday evening. As far as the archive listings go, there were ninety participants attracted to this first IBC, though a subsequent report in *Science* announcing the formation of the Society claimed there were hundred participants.

More complete details of events leading to this IBC, and in particular to the creation of the Society at Woods Hole are given in Chapter 2, as well as details of discussions during the non-scientific sessions of the IBC itself.

9.1.2 IBC 2 – Geneva Switzerland, 1949

The Second IBC was held in Geneva, Switzerland, from August 30 to September 2, 1949, immediately preceding the September 5–10, 1949 ISI Session in Berne Switzerland. The prime motivator behind this conference was Arthur Linder who was the official Conference Organizer (Linder became the Society's second President). In many ways, this IBC was a departure from the first IBC at Woods Hole (of necessity), and in many ways it featured a format

[1]The genetics work was primarily in crop and animal sciences areas, though not entirely so, rather than the microarray analyses that came to dominate in the 1990s.

FIRST INTERNATIONAL BIOMETRIC CONFERENCE

Marine Biological Laboratory
Woods Hole, Massachusetts

September 5-6, 1947.

The mathematical and statistical problems of quantitative biology have led to major improvements in scientific method, not only in biology but also in many non-biological fields. This has been achieved through the contributions of biologists, mathematicians and statisticians in many parts of the world. A primary objective of the present conference is to promote international collaboration in biometry and thereby the advancement of quantitative biological science. The role of biometry includes the formulation of quantitative hypotheses, the design of investigations for testing hypotheses and the critical evaluation of biological data by means of effective mathematical and statistical techniques.

The First International Biometric Conference has been planned at this time because a number of biometricians and statisticians from other countries are in America for the International Statistical Conferences in Washington, D. C., which continue until September 18. Moreover, many American scientists with similar interests are in this region to attend meetings of the Society for the Study of Development and Growth in Storrs, Conn., on August 26-29, of the Institute of Mathematical Statistics in New Haven on Sept. 2-4 and of the Society of General Physiologists in Woods Hole on Sept. 7-9. The Conference has been arranged by an Organizing Committee in collaboration with the National Research Council, the Biometrics Section of the American Statistical Association and the Marine Biological Laboratory, our host institution.

All times given in this program are Eastern Daylight Saving Time, which is one hour faster than Eastern Standard or Railroad Time.

page 1

PROGRAM

Friday, September 5

9:30 A.M. OPENING SESSION
Welcome to the Marine Biological Laboratory, CHARLES L. PACKARD, Director
Election of permanent officers of the Conference

10:00 A.M. QUANTITATIVE GENETICS
Chairman: A. F. BLAKESLEE, Smith College
"A Quantitative Theory of Genetic Recombination", R. A. FISHER, Cambridge University
Discussion opened by D. G. CATCHESIDE, Cambridge University

2:00 P.M. RECENT BIOMETRIC DEVELOPMENTS OVERSEAS
Chairman: E. B. WILSON, Harvard University
Informal reports by G. RASCH, Copenhagen;
O. TEDIN, Svalof;
R. C. BOSE, Calcutta
and others.

Saturday, September 6

10:00 A.M. INTERNATIONAL COOPERATION IN BIOMETRICS
Chairman: C. E. DIEULEFAIT, Rosario, Argentina
General Discussion

2:00 P.M. QUANTITATIVE GROWTH
Chairman: LESLIE F. NIMS, Brookhaven Laboratories
"La Relation d'Allométrie, sa Signification Statistique et sa Logique"
G. TEISSIER, Centre National de la Recherche Scientifique, Paris.
Discussion opened by JACQUES MONOD, Institute Pasteur, Paris.

page 2

General Information

Registration will be in the entrance hall of the Marine Biological Laboratory. All those attending the meetings are requested to register immediately upon arrival whether or not they are rooming in the dormitories. There is no registration fee.

Rooms in dormitories of the Marine Biological Laboratory will be assigned at the time of registration, in so far as they are available. The charge is $1.50 per night per person. In addition to the accommodations offered by the Laboratory, information will be at hand as to rooms elsewhere in the village.

Meals will be served at the "Mess" for $2.00 per day per person. Arrangements must be made at the time of registration. Meal hours are as follows: Breakfast from 7:30 to 8:30 A.M.; luncheon from 12:30 to 1:30 P.M. and dinner from 6:00 to 7:00 P.M.

Clam Bake: A clam bake will be held on Friday evening at 6:00 P.M. at Gifford's, Little River Road, Cotuit, about 40 minutes drive from the Laboratory. Tickets at $3.15 can be purchased at the time of registration and not later than noon on Friday. Transportation will be arranged. An adjustment in the charge for meals at the "Mess" will be made for those attending the clam bake.

Group Photograph: A group photograph will be taken in front of the Marine Biological Laboratory on Friday at noon, immediately following the morning session.

Swimming: The Marine Biological Laboratory beach is about 5 minutes walk from the Laboratory on Buzzard's Bay. Bathers should dress in their rooms and walk to the beach.

Travel Information (Daylight Saving Time)

The "Day Cape Codder" leaving New York at 9:20 A.M. and New Haven at 11:00 is due in Woods Hole at 3:25 P.M. On Thursday, September 4 only, train No. 24 leaving New York at 4:00 P.M. and New Haven at 5:30 P.M. will be met in Providence at 8:00 P.M. by a chartered bus going directly to Woods Hole. A train leaving Woods Hole at 6:35 P.M. on Saturday, September 6, reaches Boston (South Station) at 9:00 P.M. Train 173 leaving Boston (South Station) at 11:00 P.M. reaches Washington the next morning. On Sunday, September 7, trains for New York leave Woods Hole at 9:15 A.M. and at 5:30 P.M., and are due in Grand Central Station at 3:05 and at 11:15 P.M. respectively. Sunday trains for Boston leave Woods Hole at 9:35 A.M., 5:45 and 6:00 P.M. and reach South Station about two hours later.

page 3

By automobile, Woods Hole can be reached from New Haven via Middletown and Willimantic (Route 6 Alt. and 6) to Providence (101 miles) and thence via Taunton and Buzzard's Bay (Routes 44 and 28) to Woods Hole (66 miles). From Boston, the distance to Woods Hole via Plymouth (Routes 3, 6 and 28) is 80 miles.

Institutions in Woods Hole

The Marine Biological Laboratory is an independent organization under the management of biologists. It was founded in 1888 to provide a summer research station for biologists and an opportunity for students to study marine forms at first hand and receive training in scientific investigation. It is now open throughout the year and can provide research facilities for 375 investigators and their assistants. It admits about 130 students each year to its summer courses. Living material for research is furnished by the Supply Department and the Library is exceptionally complete in biological literature. Trips of inspection can be arranged at practically any time. Dr. Charles Packard is the director.

The Woods Hole Station of the U. S. Fish and Wildlife Service was established several years before the Marine Biological Laboratory. It is concerned with research on fish and shore life, especially shell fish, and now includes an aquarium and laboratories. It is open daily until 8:00 P.M. The director is Dr. P. S. Galtsoff.

The Oceanographic Institution is privately endowed and was founded in 1931 for research in oceanography. It has a permanent staff of investigators, many of whom are working on projects of the Navy. Visits can be arranged for especially interested persons. The director is Dr. Alfred C. Redfield.

Organizing Committee

C. I. BLISS (Chairman), Connecticut Agricultural Experiment Station
A. F. BLAKESLEE, Smith College
W. G. COCHRAN, North Carolina State College
E. J. DEBEER, The Wellcome Research Laboratories
H. K. HARTLINE, University of Pennsylvania School of Medicine
H. W. NORTON, U. S. Weather Bureau
N. RASHEVSKY, University of Chicago
E. W. SINNOTT, Yale University
J. W. TUKEY, Princeton University
JOHN VON NEUMANN, Institute for Advanced Study
E. B. WILSON, Harvard University

We are indebted to the Joint Arrangements Committee of the International Statistical Conferences for their cooperation and to the Rockefeller Foundation for financial aid.

page 4

FIGURE 9.1
Flyer announcement of IBC 1, with conference arrangement details.

that carried over to future IBCs. At the outset, Council approved that there be a conference committee consisting of the Society officers, representatives from Council (appointed by the President), representatives from the regions (appointed by its Vice-President), representatives from the host country and/or institution, representatives of affiliated international organizations, and delegates named by the recognized National Academy of Sciences of each participating country. Adolphe Franceschetti (1896–1968), a clinical ophthalmologist and geneticist at the University of Geneva, was the conference president; he is remembered for his delivery of his presidential address in four languages!

There were letters back and forth, primarily between Bliss and Linder, about potential speakers. They would settle on one speaker to find that he could not attend, so the discussion continued until another potential speaker was invited, and so on. The reasons given for non-acceptance of invitations ran the gamut of human emotions and proclivities – some are moving house, some are just too busy, some have no funds, one just lost his wife (which did solicit condolences), one is "buried" in the Amazon Valley. Bliss, on the other hand, was the one who wrote the letters inviting potential participants to speak. In a moment of levity, they remind each other that if a certain someone (named in the archives) accepts to speak, then strict adherence to time limits have to be enforced "remembrance of his talk at Woods Hole is invoked" (or words to that effect). A concern for a balance of speakers coming from different regions pervaded. Desires to have a range of emerging and important topics ran through the communications. Linder consulted his full conference committee in a ballot covering topics and speakers and other important aspects. However, Linder personally consulted President Ronald Fisher directly, though Fisher was a member of the committee by virtue of being a Society officer.

In recognition that the conference was being held in a French-speaking area and that Maurice Fréchet (Vice-President, i.e., "President" of the French Region) had so requested, Bliss recommended to Linder that both English and French be used as official languages in the scientific program; see Figure 9.2 for the bilingual cover page and Figure 9.3 for the bilingual content pages. This brought in questions of having translators on site. Also, the Society had by now become the Section of Biometry of the International Union of Biological Sciences (IUBS). As part of the effort to join IUBS, the Society had advocated a role to improve the teaching of Biometry; this was reflected in the teaching and education session on Friday morning. The final scientific session was for contributed papers (only one per contributor who must be in attendance to deliver the same). A business meeting was scheduled for the late afternoon of the fourth and final day. This comparatively very limited time was in sharp contrast to the dominance of what were in fact Society organizational meetings on both days of the inaugural IBC of Woods Hole, two years earlier.

This IBC began the tradition that a day be set aside for a so-called social event. This was a train trip to Berne in 1949, which was a particularly

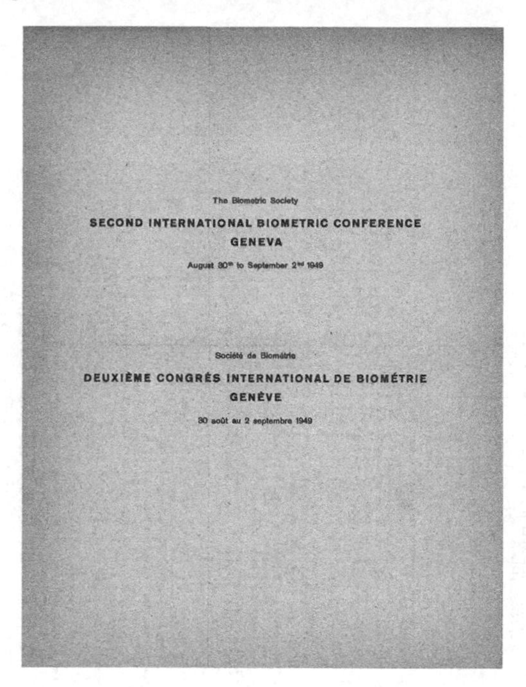

The Biometric Society

SECOND INTERNATIONAL BIOMETRIC CONFERENCE

GENEVA

August 30th to September 2nd 1949

Société de Biométrie

DEUXIÈME CONGRÉS INTERNATIONAL DE BIOMÉTRIE

GENÈVE

30 août au 2 septembre 1949

FIGURE 9.2
Scientific program cover page – IBC 2, August 30 to September 2, 1949, Geneva.

attractive option for the approximately fifty participants who planned to stay for the upcoming ISI meetings.

Archival materials dealt with the suitability of various accommodations and their costs; ultimately, hotel charges including breakfast were in the range of seven to eighteen Swiss francs per day depending on the type of hotel, meals were in the order of four plus francs. Some participants requested accommodation in the same hotels as "their friends"; Bliss himself wanted to be with other officers so as to facilitate "business" discussions, and so on. Linder had his hands full accommodating these requests. Other materials concerned travel options including charter flights (the ultimate choice, at $360 from North America to Europe). As an interesting aside in today's airline safety records, some flight details for 1949 considered safety records (one airline had "no fatalities in 150,000,000 passenger miles," for another it was 300,000,000 miles), others mentioned there would likely be no stewardess (and hence "little if any tipping") and that flying time was fourteen hours. Alternatively, travel could be by steamship passages (but it was important to book early, as these had been disrupted and affected by an internal political situation in Venezuela[!]); if traveling by boat, bicycles could be sent free on the train ticket to the boat. Arrangements for buying on the "bike abroad plan" were also detailed. The ship accommodations were deemed "austere."

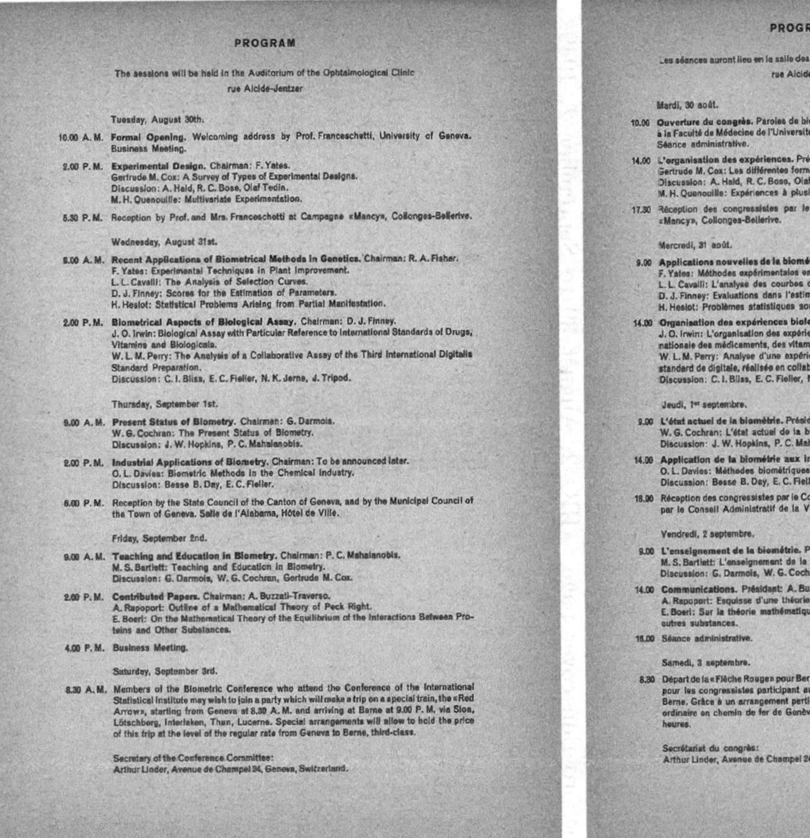

PROGRAM

The sessions will be held in the Auditorium of the Ophtalmological Clinic
rue Alcide-Jentzer

Tuesday, August 30th.

10.00 A.M. **Formal Opening.** Welcoming address by Prof. Franceschetti, University of Geneva.
Business Meeting.

2.00 P.M. **Experimental Design.** Chairman: F. Yates.
Gertrude M. Cox: A Survey of Types of Experimental Designs.
Discussion: A. Hald, R. C. Bose, Olaf Tedin.
M. H. Quenouille: Multivariate Experimentation.

5.30 P.M. Reception by Prof. and Mrs. Franceschetti at Campagne «Mancy», Collonges-Bellerive.

Wednesday, August 31st.

9.00 A.M. **Recent Applications of Biometrical Methods in Genetics.** Chairman: R. A. Fisher.
F. Yates: Experimental Techniques in Plant Improvement.
L. L. Cavalli: The Analysis of Selection Curves.
D. J. Finney: Scores for the Estimation of Parameters.
H. Heslot: Statistical Problems Arising from Partial Manifestation.

2.00 P.M. **Biometrical Aspects of Biological Assay.** Chairman: D. J. Finney.
J. O. Irwin: Biological Assay with Particular Reference to International Standards of Drugs, Vitamins and Biologicals.
W. L. M. Perry: The Analysis of a Collaborative Assay of the Third International Digitalis Standard Preparation.
Discussion: C. I. Bliss, E. C. Fieller, N. K. Jerne, J. Tripod.

Thursday, September 1st.

9.00 A.M. **Present Status of Biometry.** Chairman: G. Darmois.
W. G. Cochran: The Present Status of Biometry.
Discussion: J. W. Hopkins, P. C. Mahalanobis.

2.00 P.M. **Industrial Applications of Biometry.** Chairman: To be announced later.
O. L. Davies: Biometric Methods in the Chemical Industry.
Discussion: Besse B. Day, E. C. Fieller.

6.00 P.M. Reception by the State Council of the Canton of Geneva, and by the Municipal Council of the Town of Geneva. Salle de l'Alabama, Hôtel de Ville.

Friday, September 2nd.

9.00 A.M. **Teaching and Education in Biometry.** Chairman: P. C. Mahalanobis.
M. S. Bartlett: Teaching and Education in Biometry.
Discussion: G. Darmois, W. G. Cochran, Gertrude M. Cox.

2.00 P.M. **Contributed Papers.** Chairman: A. Buzzati-Traverso.
A. Rapoport: Outline of a Mathematical Theory of Peck Right.
E. Boeri: On the Mathematical Theory of the Equilibrium of the Interactions Between Proteins and Other Substances.

4.00 P.M. Business Meeting.

Saturday, September 3rd.

8.30 A.M. Members of the Biometric Conference who attend the Conference of the International Statistical Institute may wish to join a party which will make a trip on a special train, the «Red Arrow», starting from Geneva at 8.30 A.M. and arriving at Berne at 9.00 P.M. via Sion, Lötschberg, Interlaken, Thun, Lucerne. Special arrangements will allow to hold the price of this trip at the level of the regular rate from Geneva to Berne, third-class.

Secretary of the Conference Committee:
Arthur Linder, Avenue de Champel 24, Geneva, Switzerland.

PROGRAMME

Les séances auront lieu en la salle des cours de la Clinique ophtalmologique
rue Alcide-Jentzer.

Mardi, 30 août.

10.00 **Ouverture du congrès.** Paroles de bienvenue par le docteur A. Franceschetti, professeur à la Faculté de Médecine de l'Université de Genève.
Séance administrative.

14.00 **L'organisation des expériences.** Président: F. Yates.
Gertrude M. Cox: Les différentes formes de l'organisation des expériences.
Discussion: A. Hald, R. C. Bose, Olaf Tedin.
M. H. Quenouille: Expériences à plusieurs variables.

17.30 Réception des congressistes par le professeur et Mme A. Franceschetti. Campagne «Mancy», Collonges-Bellerive.

Mercredi, 31 août.

9.00 **Applications nouvelles de la biométrie en génétique.** Président: R. A. Fisher.
F. Yates: Méthodes expérimentales en vue d'améliorer les plantes domestiques.
L. L. Cavalli: L'analyse des courbes de sélection.
D. J. Finney: Evaluations dans l'estimation des paramètres.
H. Heslot: Problèmes statistiques soulevés par une manifestation incomplète.

14.00 **Organisation des expériences biologiques.** Président: D. J. Finney.
J. O. Irwin: L'organisation des expériences biologiques en vue de la standardisation internationale des médicaments, des vitamines et autres substances à action biologique.
W. L. M. Perry: Analyse d'une expérience sur la troisième préparation internationale du standard de digitale, réalisée en collaboration.
Discussion: C. I. Bliss, E. C. Fieller, N. K. Jerne, J. Tripod.

Jeudi, 1er septembre.

9.00 **L'état actuel de la biométrie.** Président: G. Darmois.
W. G. Cochran: L'état actuel de la biométrie.
Discussion: J. W. Hopkins, P. C. Mahalanobis.

14.00 **Application de la biométrie aux industries.** Le président sera désigné ultérieurement.
O. L. Davies: Méthodes biométriques dans l'industrie chimique.
Discussion: Besse B. Day, E. C. Fieller.

18.00 Réception des congressistes par le Conseil d'Etat de la République et Canton de Genève et par le Conseil Administratif de la Ville de Genève. Salle de l'Alabama, Hôtel de Ville.

Vendredi, 2 septembre.

9.00 **L'enseignement de la biométrie.** Président: P. C. Mahalanobis.
M. S. Bartlett: L'enseignement de la biométrie.
Discussion: G. Darmois, W. G. Cochran, Gertrude M. Cox.

14.00 **Communications.** Président: A. Buzzati-Traverso.
A. Rapoport: Esquisse d'une théorie mathématique du droit de becqueter.
E. Boeri: Sur la théorie mathématique de l'équilibre des interactions entre protéines et autres substances.

16.00 Séance administrative.

Samedi, 3 septembre.

8.30 Départ de la «Flèche Rouge» pour Berne, via Sion, Lötschberg, Interlaken, Thoune, Lucerne, pour les congressistes participant au congrès de l'Institut International de Statistique à Berne. Grâce à un arrangement particulier, le prix de ce voyage ne dépassera pas le prix ordinaire en chemin de fer de Genève à Berne en troisième classe. Arrivée à Berne à 21 heures.

Secrétariat du congrès:
Arthur Linder, Avenue de Champel 24, Genève.

FIGURE 9.3
Scientific program – IBC 2, August 30 to September 2, 1949, Geneva.

There were ninety-nine listed participants. A photo showing those seventy-six attendees who responded to the call for a photo is displayed in Figure 9.4.

Linder successfully obtained funding from insurance companies. By April 1949, he already had received what was deemed the considerable amount of 5000 Swiss francs, so that Linder asked Bliss that a small subcommittee for Finance be appointed to act as advisory to him in the disbursement of these monies; he suggested Society Treasurer John Hopkins, Secretary Bliss and himself. The Conference Proceedings were published in the March, June, and September 1950 issues of *Biometrics* with funding ($1000) from the United Nations Educational, Scientific and Cultural Organization (UNESCO); and later re-published as a booklet. A resourceful Linder arranged the entire conference "without drawing upon Society funds." He was duly complimented by both Fisher and Bliss.

9.1.3 IBC 3 – Bellagio Italy 1953

There had been very considerable energy expended by the Society to hold a conference in India in conjunction with the December 1951 ISI Session held in Delhi; this was not to be, though there was a Symposium (see Section 9.2.1). Even before this Symposium occurred, Bliss wrote to Adriano Buzzati-Traverso about having an IBC in Italy somewhere adjacent in time to an already announced major genetics conference. Buzzati-Traverso thought the idea was good, though he himself would be too heavily involved in the organization of that genetics conference but suggested Luigi Luca Cavalli-Sforza (who was reportedly "multilingual to a superb degree") be contacted for any IBC. Both of these Italian geneticists had been prominent in the early formation efforts of the Italian Region of the Society (see Chapter 5, Section 5.4).

Thus, after a brief look at other possibilities, it was determined that the Third IBC be held in Bellagio on Lake Como from September 1–5, 1953, immediately following an International Genetics Congress (August 24–31) also in Bellagio and immediately preceding the ISI Session in Rome (September 6–12, 1953); there had been not-a-little consternation judging by the considerable flurry of letters in mid-1952 between Cavalli-Sforza, Bliss and the ISI Director, when ISI seemingly wanted to change its dates. The Conference Secretary was Cavalli-Sforza (later Society President 1966–1967). Since many of those involved with the organization of the genetics conference were Society members and also involved with the IBC and since the on-the-ground organization of the former would be in place already for an IBC, this was deemed a most desirable choice.

An emphasis on genetics pervaded the IBC scientific program. Other sessions dealt with methodology in biometry, biometric methods in agriculture, functional relations in experimentation, and industrial applications of biometry. There was also a session on training in biometry (under the auspices of William G Cochran, with a repeat version by Cochran at

SECOND INTERNATIONAL BIOMETRIC CONFERENCE – GENEVA, AUGUST 31, 1949
DEUXIEME CONGRÈS INTERNATIONAL DE BIOMETRIE – GENEVE, 31 AOUT 1949

(a)

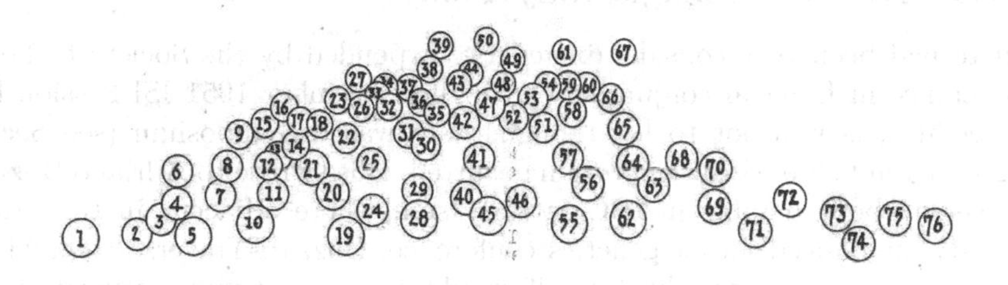

1 E.C. Fieller	26 M. Schützènberger	51 W.L.M. Perry
2 E.F. Drion	27 V. Tonolli	52 F.B. Leech
3 A. Linder	28 G. Rasch	53 P.S. Øestergaard
4 D. Klein	29 F. Brambilla	54 N.K. Jerne
5 K.L. Smith	3o Frank Yates	55 R.A. Fisher
6 W. Keiser	31 M.H. Quenouille	56 G. Goudswaard
7 Mrs. W.H. Norton	32 H. Åstrand	57 J. Reboul
8 Stuart A. Rice	33 A. Lubin	58 J.B.S. Haldane
9 A. Rapoport	34 R. Jardine	59 S. Rosin
1o J.M. Vincent	35 Besse B. Day	6o Ole U. Maaløe
11 W.G. Cochran	36 H. Florin	61 Ch. A.G. Nass
12 R.T.H. Beverton	37 L.L. Cavalli	62 G. Darmois
13 Mrs. A. Rapoport	38 R. Scossiroli ne	63 M.S. Bartlett
14 T. Dalling	39 N. Visconti di Modro-	64 A.A. Weber
15 A. Buzzati-Traverso	4o Sir Percival Hartley	65 Leo Törnqvist
16 Eric Reeve	41 D. van Dantzig	66 M.W. Bentzon
17 Livia Pirocchi	42 M.J.R. Healy	67 R. Borth
18 R. Féron	43 D.J. Finney	68 J.M. Legay
19 K.B. Madhava	44 W. Wegmüller	69 S. Agapitidès
2o E. van der Laan	45 Chester I. Bliss	7o E. Morice
21 A. Franceschetti	46 S.T. Bok	71 Corrado Gini
22 E. Dieulefait	47 G. Pompilj	72 H.J. Stutvoet
23 Mme. E. Defrise	48 J. Tripod	73 D. Schwartz
24 Gertrude M. Cox	49 J. Jenny	74 A. Vessereau
25 J.W. Hopkins	5o L. Martin	75 O.L. Davies
		76 R. Karschon

(b)

FIGURE 9.4
IBC 2 attendees – photo (a) and names (b) – Geneva 1949.

the ISI Session) and a sub-committee on standardization of symbols (chaired by Frank Yates, also to report at ISI), both contemporary topics of the day. The standardization of symbols issue had lurked in the background for consideration at IBC 2 but never eventuated. Here at IBC 3, Yates reported and recommended that the Society discharge this committee as the issue was being taken up by ISI and other international entities. As for IBC 2, there was an extensive correspondence between Bliss and the Conference Secretary (this time Cavalli-Sforza) especially about session topics and speakers. There were also several letters from members asking if they could contribute a paper – Contributed Sessions did become part of the scientific program.

President Georges Darmois delivered a Presidential Address in French which was published in *Biometrics* (Darmois, 1953), followed by an English summary from Cavalli-Sforza (not published). As for the Geneva IBC, a booklet containing abstracts and papers (from *Biometrics*) was published, and contained details of the scientific program, a Report on the Committee on the Teaching of Biometry, and a list of the 125 registrants from 24 countries. Instead of translations after a presentation as in Geneva, this time there would be simultaneous translations (though more expensive, but considered more useful). In addition to the scientific program, there were five Exhibits including one from Dufrenoy and Goyan on "A graphical calculator for statistical analysis (illustrated by Scwhartz)." Very explicit instructions to contributors on all aspects of preparing papers, including any diagrams (see, e.g., Figure 9.5) were spelled out in detail. Bliss, in cahoots with Cavalli-Sforza, hoped that authors of *Proceedings* papers would have them submitted to *Biometrics* in time so that galleys could be proofed at Bellagio itself; Editor Gertrude Cox reacted "Some dreaming you are doing." The IUBS gave funds (about $600) for publication of the *Proceedings*, as well as some monies for travelling expenses "to ease the participation of junior specialists from war damaged counties."

An exhibit of books from publishers, set up next to the conference Secretarial Bureau, was a Cavalli-Sforza innovation prompted by an overture from a publisher. Without missing a beat, Bliss immediately wrote (it is now late July 1953) to other book publishers inviting them to be part of this exhibit – only weeks away from the conference. Bliss was "amazed" to read Wiley's

2. Diagrams should be drawn in black ink on white paper or board, or on graph paper with faint blue lines. They should be about twice the size of the finished block, which cannot exceed $4\frac{1}{2}$ x 7 inches, each diagram should carry on the back the author's name, the title of the paper and the number in the diagram. Letters, numbers, etc. on diagrams should be indicated lightly in pencil. Legends should be typed separately and numbered to correspond with the relevant diagrams. Copies of the diagrams should be submitted where possible.

FIGURE 9.5
IBC 3 – instructions to contributors for diagrams.

reply that they did not have any books pertinent to conference participants; Bliss quickly replied that Wiley authors such as Fisher, Cochran, Cox, Rao, and probably others, were IBC participants. Wiley came to the show, bringing mostly the same books they planned to exhibit at the ISI Session. The difficult point of charges for the exhibitions arose after the event; these were mostly resolved by donating the books to the Italian Region.

Accommodations ranged from "Deluxe" with bath at $10 per day to a "Good Hotel" without bath at $3.50 per day, including breakfast and one meal; members were advised that bookings should be made by February 1, as Lake Como was a popular resort. For the first time, the Conference had a registration requirement (at $6, and $4 for accompanying persons) which included the Banquet. A new addition was a Ladies Programme, with social events planned for each of the four conference days, including several trips to some of the surroundings of Bellagio.

9.1.4 IBC 4 – Ottawa Canada 1958

Because IUBS was no longer providing funding for established conferences, in this case IBCs, the hoped-for-IBC in Brazil was instead a Symposium; see Section 9.2.2. Discussions therefore began for the Fourth IBC, which became officially the "IVth International Conference on Biometry and Symposium on Biometrical Genetics," held in Ottawa, Canada August 28 to September 2, 1958. This was preceded by an International Genetics Conference in Montreal (August 20–27, 1958). The Conference Organizing President was John W Hopkins and Local Arrangements Chair was Gail B Oakland. The Symposium component allowed for IUBS funding support.

The first day was devoted to Biometrical Genetics (sessions (1)–(3) organized by the Genetics Symposium Committee chaired by Oscar Kempthorne, ENAR). The working languages were French and English, with Abstracts in both languages where possible. The archives give scant attention to the details of this conference, except for one brief tentative program. As an aside, it should be said that, based on an archive letter where it is reported that Hopkins is now feeling better from being indisposed and his travails when trying to serve as *Biometrics* Editor (see Chapter 7, Section 7.4), Hopkins was clearly heading into a time of serious illness, and so understandably might not have been able to assemble the records for the Society he might well have provided otherwise. However, the actual scientific program did appear in *Biometrics*, from which we learn that scientific sessions dealt with (1) theoretical genetics; (2) design of experiments; (3) experimental results; (4) multivariate analysis (organized by Evan J Williams, AR); (5) chi-square and other biometric techniques (William G Cochran, ENAR); (6) interpretation of experimental results (Theodore (Ted) Bancroft, ENAR); (7) mathematical and statistical models in biology (Léopold Martin, RBe); (8) biometry in clinical research (Donald Mainland, ENAR); and (9) ecology and animal behaviour (Douglas G Chapman, WNAR). The Conference opened with an Address from Society President Cyril Harold Goulden (ENAR).

Most of the archival records dealt with funding requests and amounts received with quite some detail about how much could be spent to cover what costs, and the like. There were also applications from participants for travel support, again with detailed requests. Funding grants came from IUBS ($1000, for travel through the International Council of Scientific Unions, ICSU) and the National Science Foundation (NSF, $2500, obtained by the sponsoring host ENAR's President Boyd Harshbarger) along with the Canadian National Research Council ($6000). The conference Financial Executive consisted of President Goulden, Treasurer Allyn Kimball, Hopkins and Oakland. In what was deemed to be an extraordinary move, ICSU accountant paid C R Rao a personal check ahead of time; but insisted that (as per their policy) other funds for travel were for reimbursements to be paid after the conference as "it is not desirable to pay a grant to someone who cheerfully accepts the money and does not attend ... [and that] before pay[ment] ... a list of actual participants be sent [to the accountant]." A portion ($700) of the cost of the *Proceedings* was also covered by the ICSU grant. The Biometrical Genetics component of the *Proceedings* was published by Pergamon Press, while *Biometrics* published the IBC part. There were 207 paid-up participants.

Registration was $10 for the US and Canadian participants and £3 sterling for others, while lodging was $8.00 per night. Though not covered in the archival materials, the Report in *Biometrics* did tell of a women's social program and a Sunday tour of the Gatineau National Park for participants; and that John Tukey's Banquet Speech produced the epigram "Problems we do not understand, we call multivariate." [Pure Tukey!]

9.1.5 IBC 5 – Cambridge United Kingdom, 1963

The Fifth IBC was held at Cambridge, England, September 10–14, 1963, following the XI International Genetics Conference (September 1–10, 1963) in The Hague. For the first time, a local organizing committee was appointed to remove the burden from the officers. Michael J R Healy was the Organizing/General Secretary, and R Colin Campbell was Local Arrangements Secretary with John G Rowell as Conference Treasurer. With Fisher's death (on July 29, 1962), the IBC became much more than a normal IBC; it was also a celebration of Fisher himself. In his roles as Society Secretary (1956–1962) and IBC Organizer, it is not surprising to learn that Healy kept voluminous records for the archives, including abstracts and paper drafts, and even the train schedule to and from London and Cambridge. Considerable prior work allowed for a well planned conference. Indeed, in anticipation of the September 28, 1961 formal invitation from the British Region (BR), Campbell wrote to Healy on September 22, 1961, with extensive details of plans already formulated. Immediately, in November 1961, Treasurer Marvin Kastenbaum sent Campbell $500 to cover early expenses.

Members intending/desiring to present a paper at IBC 5 were to submit an Abstract by September 20, 1962. Given the anticipated high attendance, it was understood that some papers would be by title only, a feature not always appreciated by potential participants. Healy grouped these abstracts under various subject headings, with that subject's session organizer making the selection of which papers to accept for oral presentation and which for title only. Since the papers were to be part of the *Proceedings* (to be ready ahead of the Conference), a deadline for submission of the manuscripts was set as mid-May but later advanced to April 1, 1963. Non-English-speaking authors were to provide a 500-word summary for translation into English; Healy who was fluent in both English and French would see to the translation if necessary. Healy had set some quite specific guidelines as to how the papers themselves were to be written/formulated. These guidelines elicited a colorful response from Norman T J Bailey asking about writing mathematical formulae by hand; Bailey could "see this is advisable for certain primitive societies, but he has a proper mathematical typewriter, and is reluctant to return to a goose-quill ... ." Healy agreed "if they come out legible when xeroxed."

By November 1962, letters accepting papers/abstracts for oral presentation were going out along with rules pertaining to the submission of the full text. Other letters politely declined submissions with the added hope that the paper be submitted somewhere. Some were declined because the topic did not seem to fit into the overall theme, and others because there were simply too many for a given session to have all of them included; these folk were issued invitations to present their work by title. Occasionally, a "declined" for oral presentation paper author replied unhappily, at which stage the session organizer would engage Healy to help heal the wounds. Some of these disappointed authors wrote voicing their feelings while some asked about having parallel sessions (which minority view option had earlier been dismissed as in previous IBCs). [All in a day's work when organizing meetings! These particular archive records reveal it all.] Though the scientific program had all but been completed by 1962, letters requesting the writer be included in the program still arrived to which Healy sometimes offered the option of presentation by title while rejecting others; on March 1 1963, Healy thinks it is now too late even for that option, though some requests still came later. This was clearly going to be a popular well-attended IBC attracting a lot of interest. In his final Report, Healy elaborated somewhat on the issue of parallel sessions (by implication at the time these would be contributed papers), reporting that there were strong feelings expressed both for and against, concluding that the Society "must decide" if "simultaneous sessions" were to be part of future conferences.

Fisher died at the end of July 1962 (29th). Yet, within a month, Conference Organizer Healy had written to previously invited session organizers proposing that each session start with a talk focussing on Fisher's contribution to the subject area of that session; all had agreed. Most "knew" immediately what this would entail, though occasionally there was a gentle nudge from Healy himself. Unsurprisingly, there were many communications dealing with this aspect of the scientific program. Over the ensuing months, Healy continued to issue invitations to other potential key presenters for these sessions. It seemed that very few declined.

Healy was very well organized (in fact the archives contained many communications congratulating him on his excellent organization of the Conference). In June 1963, he sent all potential participants a twelve-page document, detailing the outlines and structure of the conference, a preliminary scientific program (a portion, reproduced in Table 9.1, highlights the nature of the Fisher celebration that permeated through the scientific program), a social program for Conference members and another social program for "Family members" (note "family" rather than wives of earlier conferences), rules for Gonville and Caius College occupants (such as the gates closed at 11 P.M.) – all the conference go-er needed to know. It is also interesting to note that the Conference accommodation was in this college, Fisher's Cambridge college, though this had been arranged before Fisher's passing, possibly by Fisher himself (the archives do not give us that detail); however, the Local Organizing Secretary Campbell was a Caius man.

The Conference itself had a grand opening at the Monday reception hosted by the British Region, with the Right Honourable the Viscount Hailsham Lord President of the Council and British Minister of Science (introduced by British Regional President John Gaddum), the Right Worshipful the Mayor of Cambridge (J B Collins), the Registry of the University of Cambridge, and the Master of Gonville and Caius College. Lord Hailsham's Address and Society President Bliss' reply are published in *Biometrics* (see Campbell and Healy, 1963). The spirit of Fisher permeated everywhere. The pomp and ceremony as befits a royal occasion was evident throughout. The schedule of formal dinners for officers prompted Bliss to write to Healy asking "what should I wear as [Society] President, tuxedo, academic robe, or just ordinary day clothes?"

Also, by June 1963, Healy advised that an approximately 500-page volume of preprints was in press and to be mailed to European registrants before the Conference; though costly largely because of the number and length of papers received, this initiative was well received. There was no formal proceedings outlet for papers in general, though abstracts were to appear in *Biometrics* in late 1964. Ultimately, the highlighted Fisher papers were published in a special June 1964 "In Memoriam Ronald Aylmer Fisher 1890–1962" volume of *Biometrics*. This Volume started with a message from President David Finney, followed by a biography by Mahalanobis (1964). This Bibliography of Fisher's contributions to biometrics included three papers for reproduction, one of which was Fisher's (1948) brilliant address to British members at the formation of the British Region.

There were 299 registrants from 31 countries. Along with support of £356 from IUBS, many companies also contributed funds totaling £1216 (at the time £1 sterling was equivalent to US$2.80). This included £100 from Spillers (Flour Millers and Feeding Stuffs Manufacturers) specifically to support Secretary Henri Le Roy's attendance; Spillers also made arrangements

TABLE 9.1
Portion of Scientific Program – IBC 5, September 10–14, 1963, Cambridge UK – Featuring the Celebration of R. A. Fisher's Contributions

Session I: Tuesday September 10 morning	
	Chairman: J. H. Gaddum, Organizer: P. D. Oldham
	Medicine and Bioassay
C. I. Bliss	**The work of R. A. Fisher in medicine and bioassay**
Others[a]:	Goldberg, Matthews, Weiss and Zelen, Sheps and Perrin, Bailey, Zippin, Rustagi
Session II: Tuesday September 10 afternoon	
	Chairman: M. W. Bentzon, Organizer: R. E. Blackith
	Multivariate Analysis
C. R. Rao	**R. A. Fisher – the architect of multivariate analysis**
Others[a]:	Williams, Dagnelie, Geisser, Oldham and Rossiter, Brieger and Vencovsky and Packer, Rouvier and Vissac, Edwards and Cavalli-Sforza, Jeffers
Session III: Wednesday September 11 morning	
	Chairman: J. O. Irwin, Organizer: L. Martin
	Mathematical Models in Biology
C. B. Williams	**R. A. Fisher from a biologist's viewpoint**
Others[a]:	Turner and Monroe, Newell, Mode, Weiling, Olsen, Wiggins, Wilhelm
Session IV: Thursday September 12 morning	
	Chairman: D. R. Cox, Organizer: M. J. R. Healy
	Experimentation
F. Yates	**R. A. Fisher and the design of experiments**
Others[a]:	Draper and Stoneman, Nelder, Kemp, Harshbarger, Pearce, Freeman, Taylor, Becker and Bearse
Session V: Thursday September 12 afternoon	
	Chairman: D. J. Finney, Organizer: C. W. Dunnett
	Screening and Selection
Speakers[a]:	Davies, King, Brock and Schneider, Searle, Kempthorne, Curnow
Session VI: Friday September 13 morning	
	Chairman: Helen N. Turner, Organizer: D. J. Finney
	Statistical Methods
D. J. Finney	**R. A. Fisher's contributions to biometric statistics**
Others[a]:	Juvancz and Fischer and Csaki, Bradley, Greenburg and White, Legay and Pontier, Sarhan
Session VII: Friday September 13 afternoon	
	Organizer: H. F. Robinson
	Statistical Methods
K. Mather	**R. A. Fisher's work in genetics**
Others[a]:	Binet, Le Roy, Bohidar, Baker, Curnow, Cockerham, Hartmann, Hanson, Robertson

([a]Some by title only.)

for Le Roy to be met at Cambridge and to make his way to the airport. Interestingly, the budget included the cost for a chauffeur for Lord Hailsham. Housing was £2.50 or £2.80 depending on the "court" of Gonville and Caius College. The registration fee was £5.25 (and £7.37 after April 30, the first time to have early/late registration dates); a one-day registration was £2.10, or £3.15 to include the volume of preprints. Except for dinners and an excursion to Stratford-upon-Avon, social activities for Family members and Conference members were free. An NSF grant to assist with travel costs for ENAR and WNAR members was obtained by H. L (Curly) Lucas (1962 ENAR President). The IUBS subvention also included $500 in support of the Symposium on the Use of Statistics in Hospitals being held in conjunction with the ISI Session immediately following, but in Ottawa; this Symposium took the form of a jointly sponsored Society-ISI session.

A Study Conference followed the IBC. This was intended to allow biometricians, especially isolated ones, to receive advice and extended discussions on four recent topics (suggested by a survey of potential participating members), in many ways a precursor of later short courses. This proved to be a popular addition.

The 1959 Council reported that the German Region had invited the Society to the 5th IBC in Germany in 1963; there was a separate effort for the Genetics Congress also to be in Germany in 1963. Council was asked to decide but with the caveat that "it might be wise to postpone it until the political situation is a little less obscure." The Society did go to Germany but in 1970 for the Seventh IBC. Healy's caution here was not without its merits. For the Cambridge IBC, there had been letters between him and the Foreign Office regarding visa problems for German Region members living in Berlin.

In an odd twist, after the IBC, in 1964, Campbell was caught up in a dispute with two photography firms wanting him to make certain conference documents available. Finney stepped in, explaining how the release of related correspondence would be improper without approval from Council. Finney suggested that Sir John Gaddum, President of the British Region, should also be able to advise Campbell.

9.1.6 IBC 6 – Sydney Australia 1967

The Sixth IBC was hosted by the Commonwealth Scientific Industry and Research Organization (CSIRO) and held in Sydney Australia over August 20–25, 1967. The IBC was held in the brand new Wentworth Hotel in the center of downtown Sydney – rather than at a university or similarly-like research location. This IBC immediately preceded the ISI 36th Session in the same location (same hotel) from August 28 to September 7, 1967. The Conference Secretary was H. R. Webb (of CSIRO) and Australasian Regional President Evan J Williams served as Conference Chair.

Discussions as to where the next IBC, after Cambridge in 1963, should be held had begun in late 1963, with the highly visible Australasian members (and former officers, Helen Turner and Alf Cornish) offering Australia as host. Society officers looked at several options (Sydney, Belgrade, Split, Dubrovnik, ...), usually as was often the case through a prism of what selection might enhance the growth of the Society. Financial costs of traveling to Australia were major concerns. However, with the settling of the ISI Session in Sydney, progress to accepting the invitation from CSIRO readily followed.

The scientific program covered ten half-day sessions on different topics – medical and physiological biometry (2 sessions), quantitative aspects of ecology and taxonomy (2 sessions), experimental strategy (1.5 sessions), role and influence of computers in biometrical research (2 sessions), teaching mathematics and statistics for biology, and homeostatic mechanisms in biological systems (1.5 sessions). Cavalli-Sforza gave his Presidential Address at the Conference Dinner on the final evening, where he expounded on his long-running interest in teaching biometry in the schools; see Cavalli-Sforza (1968). In addition to these 52 papers, 26 were read by title only; the 95 abstracts were published in *Biometrics* in March 1968. The Conference had been opened officially at the Monday evening Reception by the Honorable Senator John G Gorton, Australian Government's Minister for Education and Science (later Prime Minister). There was a pre-conference tour of Sydney Harbour on the preceding Sunday (August 20).

Early concerns had been raised by the regional presidents of France and Belguim that presentations would be in English only. Subsequent discussions between then-President David Finney and the incoming (1966–1967) President Cavalli-Sforza decided to allow presentations in French and/or German but without translators.

Not surprisingly, given the relative isolation of Sydney and the associated considerable distances and costs to attend, many archive letters dealt with funding sources, where and who might provide funds, applications for same, and similar issues. Being back-to-back with the ISI Session helped in these endeavors. At one point, the Society Treasurer Henry Tucker increased the travel support to Secretary Henri Le Roy on the grounds that it was deemed "essential" that he, and in fact all Executive officers, be present. The final supporting funds, in addition to the Australian government's support of AU$16,000 (at the time, AU$1 ≡ US$1.12), consisted of the Society's contribution $2500, and $1500 advanced by IUBS. Monies from IUBS had a subtlety. The IUBS did not want to support an IBC per se, but did provide monies to serve as loans to help in the preparatory work, which would be converted to grants if the financial situation did not allow for a refund; the 1968 accounts showed a "return from conferences" of $1187.05. There is an excellent detailing of the income/expenses in the June 1968 issue of *Biometrics.* Importantly, the final accounting reported on a profit of about $10,470; so a huge financial success, quite unexpected given the early fears regarding costs. This surplus was distributed as repayments to the Australian government ($7646.44), IUBS ($638.68) and the Society ($1062.98). In addition, grants from the US funding agencies, administered by ASA, assisted ENAR/WNAR members.

Registration was AU$30 for members and AU$35 for non-members and AU$15 for family members. Hotel accommodation ranged from AU$5 single room without bath to AU$13.59 for a twin-bedded room with bath. There were 241 participants from 17 countries (with 20 family members), according to the official Report.

After the IBC was over, Secretary Le Roy wrote to Conference Chair Evan Williams declaring that its success was due to its good organization and the good scientific program even though the geographical isolation and the costs to attend would be difficult. Treasurer Tucker concurred with "unprecedented success ... both financially and in attendance, is entirely to the credit of the Conference Committee and especially the Conference Secretary [Webb]." As Le Roy reported to Council, it was a "great success" and the "relaxed Australian hospitality ... put Australia on the map."

9.1.7 IBC 7 – Hannover Germany 1970

The Seventh IBC was held in Hannover Germany August 16–21, 1970. The Local Arrangements Committee Secretary was Berthold Schneider who was also President (1970–1971). Given Schneider's dual roles, it was felt that a separate Scientific Program Chair was needed; Peter Armitage (BR) was thus appointed, though Schneider also served on the Program Committee. Up to this IBC, the Society officers were part of the Conference Committee along with a representative from the host region. Discussions among the officers suggested that there should now be both an organizing committee and a program committee both appointed by the President; these changes were included in the Constitution and By-laws revisions approved in 1974.

Letters as early as December 1967 conveyed the news that discussions were underway for the next IBC after Sydney. Some discussions included proposals to hold this in the US, particularly with ENAR/WNAR, to coincide with the Washington, DC ISI 1971 Session. In the end, it was 1976 when an ENAR/WNAR sponsored IBC actually occurred, in Boston; see Section 9.1.9. One concern was that Council in 1967 felt there should not be more than four years between IBCs. Also, the German Region was anxious to host an IBC; so discussions about dates followed including the possibility that the IBC be in an Oberammergau year[!]. By January 1969, the selection of Hannover for 1970 had been made.

The Conference was opened Monday morning, August 17, with Schneider's Presidential Address (in German, "Biometrie den 70er jahren", i.e., "Biometrics in the 70s"; see Schneider, 1971). The scientific program consisted of two daily half-day invited sessions except Wednesday when the afternoon was free for the organized excursions. The topics covered (1) Testing and monitoring of drugs (organizer Léopold Martin, RBe); (2) multivariate methods (Evan J Williams, AR); (3) medical documentation and computation

(G Wagner, DR); (4) contingency tables (Robin L Plackett, BR); (5) planning of experiments (S Clifford Pearce, BR); (6) biological assay (Chris L Rümke, ANed); (7) statistical methods in animal and plant genetics (Henri L Le Roy, ROeS, and Hans Rundfeldt, DR); (8) mathematical models in biology (Bernard G Greenberg, ENAR); (9) human genetics (Luigi L Cavalli-Sforza, RItl); there were five Contributed paper sessions held in parallel with the afternoon invited sessions, except for Wednesday when this was in the morning. Clearly, those who had advocated (unsuccessfully) for contributed sessions when planning the Cambridge IBC, now had their opportunity to present (as had incidentally been recommended for the future at the time, see Section 9.1.5). The official languages were English, French and German, with simultaneous instantaneous translations. The papers were not published in *Biometrics*, but the Abstracts were published in the March 1971 issue. An extensive "social programme" included a visit to the Volkswagen Works in Wolfsburg.

Schneider and the German Region were very successful at garnering local support. The town itself gave DM20000 ≡ US$11650 at the time to the Conference as did the German Region also give DM20000, along with pharmaceutical firms promising considerable contributions. With these funds, Schneider did not think there was a need to seek IUBS support, though individuals might be so helped. The Finance Committee voted to budget $2500 for travel expenses for invited speakers allotted by the Program Committee and also expressed the view that Executive Officers should attend and be supported. Funds applied/requested from the Public Health Service and the Atomic Energy Commission (administered by ASA) assisted WNAR and ENAR members.

9.1.8 IBC 8 – Constanza Romania 1974

The first IBC to be hosted by an Eastern European country was the Eighth IBC in Romania on August 25–30, 1974. In January 1973, this had been approved for Bucharest, but when President Peter Armitage and Leo C A Corsten (Program Chair, ANed) visited immediately following the Vienna 1973 ISI, they learned that a World Population Conference was to be held in Bucharest on those dates, which totally precluded other Bucharest meetings. After considering possibilities, the decision was made to switch the IBC to Constanza (Constanta) on the Black Sea.

Questions had reigned over the site of the next IBC after Hannover as finding a host site proved difficult. An invitation from Hungary was withdrawn in 1971, while the Austrian members were not prepared to host in conjunction with the ISI Session in Vienna. Brazil was also involved, but their meeting eventuated in 1979 (see Section 9.1.10). It was known ENAR planned to host in 1976, but the Society did not want to wait that long for the next Conference. Eventually, an invitation came from the President of the Romanian Academy, appreciatively accepted. The Chair of the Organizing Committee was Tiberiu Postelnicu (Romanian National Secretary).

This time, IUBS rejected the application for a subvention. Treasurer Larry Nelson was clear that funds could not be provided for all invited speakers.

9.1.9 IBC 9 – Boston USA 1976

Approved in early 1972, the second IBC to be in the USA was the Ninth IBC held in Boston Massachusetts USA on August 22–27, 1976, in conjunction with the ASA Annual Meeting. Chair of the Local Organizing Committee was Yvonne M M Bishop and the Scientific Program Committee was chaired by Edmund A Gehan (ENAR); ASA had its own separate Local and Program Committees. The local organizers designated separate chairs responsible for travel and for housing. Both ENAR and WNAR issued the invitation for this IBC.

The backdrop to the selection of Boston was the 200th Anniversary of the Declaration of Independence of the USA along with Boston's close proximity to Woods Hole where the Society was founded. Thus, the Social Program included a half-day trip to Plymouth (site of the first English pilgrim settlers) followed by a clam bake (a feature of the Woods Hole's IBC social program). In keeping with this historical note, the Welcoming Ceremony, held on the Sunday evening, featured addresses by Chester I Bliss, William G Cochran and Gertrude M Cox, all of whom played important roles at the Society's founding and all former presidents; this was followed by an Address from then President Henri L Le Roy.

The official program listed 137 sessions, of which 58 were invited sessions and 79 were contributed sessions (including three poster sessions – one from our Society and two from ASA). Untangling those invited sessions that were organized by the Society gave: (1) regression methods (organized by Ako Kudô, JR); (2) multivariate methods in biometry (Ram Gnanadesikan, ENAR); (3) design of experiments (Dieter Rasch, RGDR); (4) planning clinical trials (Marvin Zelen, ENAR); (5) interim data analysis in clinical trials (Peter Armitage, BR); (6) contingency tables (Gary G Koch, ENAR); (7) early history of biometry (Churchill Eisenhart, ENAR); (8) methodology of survival analysis (John J Gart, ENAR); (9) growth studies (Ettore Marubini, RItl); (10) stochastic processes in biology and medicine (Peter A W Lewis, WNAR); (11) statistical contributions to environmental problems (David G Hoel, ENAR); (12) mathematical models in biology and medicine (Klaus Dietz, ROeS); (13) statistical computing (Ivor S Francis, ENAR); (14) statistical methods in genetics (Robert C Elston, ENAR); (15) survey methodology (Lyle D Calvin, WNAR); (16) statistical ecology (James Mosimann, ENAR); all Society invited sessions were co-sponsored by ASA's Biometrics Section, and one was co-organized with ISI. The forty-six invited presentations and twenty-one discussants, from seventeen regions/groups across the world, brought a true international presence to the meeting. The Society's contributed paper abstracts could be in English or French and were limited to 300 words; these were organized into 23 Contributed Sessions (with 165 presentations). The *Proceedings* containing the invited papers and contributed abstracts were published as a two-volume set and distributed at the Conference; these were also available for purchase. While the published program listed 774 active participants, no distinction was made as to who might have been a Society member. It can be inferred, however, that at least 233 were Society members; officially, over 600 members were estimated to have attended.

Funds were provided by more than thirty primarily pharmaceutically oriented companies. Treasurer and Business Office Manager Nelson garnered and administered a total of $56000 as grant awards from various NIH institutes for travel expenses for 40 non-US members, with an extra $2000 being awarded directly to the Society. The daily expenses per participant were estimated at US$35. Registration was $25 for members and $30 for non-members.

9.1.10 IBC 10 – Guarujá Brazil 1979

The Tenth IBC was held in Guarujá Brazil on August 6–10, 1979, at the invitation of the Brazilian Region (RBras). The Local Organizing Chair was Décio Barbin. The Scientific Program Chair was Geoffrey H Freeman (BR), with co-chair Roland Vencovsky (RBras) handling the contributed papers.

There were thirteen invited sessions, viz., (1) exploratory data analysis (organized by Norman A Goodchild, AR); (2) fitting nonlinear models (Kei Takeuchi, JR); (3) experiments with fertilizers (Larry A Nelson, ENAR); (4) multivariate analysis of structured data (Richard Tomassone, RF); (5) linear models in agricultural data (Tadakazu Okuno, JR); (6) analysis of longitudinal data (James E Grizzle, ENAR); (7) biometrical aspects of genetics (Cedric A B Smith, BR); (8) adapting experimental designs to the needs of practical problems (Charles W Dunnett, ENAR); (9) cluster analysis (John C Gower, BR); (10) clinical trials (Edmund A Gehan, ENAR), (11) analysis of survival data (Geoffrey Berry, AR); (12) biometric problems in ecology (Jean M Legay, RF); and (13) applications to environmental health problems (Yvonne M M Bishop, ENAR). A volume of contributed abstracts was available for purchase; *Proceedings* became available in 1983.

At the August 1979 Council meeting at this Guarujá IBC, invitations to host from the French Region and the Japanese Region were accepted for 1982 and 1984, respectively, fitting in and around ISI meetings. This Council also recommended that host regions be approved at least four years and the Program Chair at least three years, in advance; and that the IBC be held in the language of the host region and in English.

The Guarujá 1979 Council meeting also approved that the Society give $10,000 in support and to pay full expenses of the President and travel expenses for the rest of the Executive Committee plus a representative from the next IBC – up to a maximum of $15,000. This was distributed as $5000 for direct expenses, and if necessary and requested $5000 as an interest-free loan. The Society was not to incur liability incurred by the host region for the meeting. In particular, in 1980, this Committee recommended that someone from the Local Organizing Committee of the succeeding IBC be in attendance at a given IBC to learn from the experiences encountered (there was always much to learn).

9.1.11 IBC 11 – Toulouse France 1982

At the instigation of Richard Tomassone, the French Region (RF) invited the Society to France for the Eleventh IBC held from September 6–11, 1982, at l'Universite Paul Sabatier Toulouse France. The Local Organizing Chair was Jacques Badia. The Scientific Program Chair was Charles W Dunnett (ENAR), with Badia handling the contributed papers. The original site was to be Paris, approved in 1980; in September 1980, we learn that Tomassone wanted to shift to Toulouse to coordinate with a COMSTAT meeting there on adjacent dates.

At Toulouse, the Invited papers program covered topics: (1) Modelisation des phenomenes biologiques; Building and fitting mathematical models to biological data (organized by Daniel L Solomon, ENAR); (2) planification et evaluation statistique d'etudes epidemiolog ques des maladies chroniques non infectieuses; design and statistical evaluation of epidemiologic studies of non-infectious, chronic diseases (Gottfried Enderlein, RGDR); (3) ecologie statistique; statistical ecology (Jean-Dominique Lebreton, RF); (4) planification experimentale; planning experiments (Peter W M John, ENAR); (5) applications bio-pharmaceutiques; bio-pharmaceutical applications (Hugo Flühler, ROeS); (6) problems en analyse d'essais clinques; issues in the analysis of clinical trials (James H Ware, ENAR); (7) avenir des biostatistiques, une profession et ou une discipline; future of biostatistics as a profession and or discipline (Samuel S Greenhouse, ENAR); (8) modeles lineaires en experimentation agronomique; formulation of linear models for agricultural experiments (Akio Kûdo, JR); (9) tables de contingences multidimensionnelles-donnees categorielles; multidimensional contingency tables-categorical data (Joseph L Lelloch, RF); (10) statistiques en amelioration genetique des animaux et des vegetaux; statistics in animal and plant breeding (William G Hill, BR); (11) processus stochastiques en biologie et medecine; stochastic processes in biology and medicine (Peter D M Macdonald, ENAR); (12) combinaison des resultats issus de plusieurs ensembles de donnees; combination of results from several data sets (Geoffrey H Freeman, BR); (13) besoins en biometrie dans tiers monde; needs of biometry in the third world (Pierre Dagnelie, RBe); (14) problemes biometriques en genetique; biometric problems in genetics (Robert N Curnow, BR); (15) calcul en statistiques-analysr des donnees, besoins en biostatistiques; statistical computing-data analysis needs in biostatistics (Anna Bartkowiak, GPol). There were two-three invited papers per session, and papers (except for session (11)) were published in a *Proceedings* distributed at the IBC.

It was during this IBC that Council decided that IBCs should be held every other year, in the even-numbered years off-set by the ISI biennial sessions held in odd-numbered years. This was a significant decision given that the IBCs had started in conjunction with corresponding ISI Sessions, and underscored the fact that the Society was now an important association in its own right and no longer felt the need to be tied to any other association for legitimacy.

9.1.12 IBC 12 – Tokyo Japan 1984

The Twelfth IBC was held in Tokyo Japan from September 2 to 8, 1984. The Chair of the Local Organizing Committee was Chikio Hayashi with Tadakazu Okuno as Conference Secretariat, and the Scientific Program Chair was Daniel L Solomon (ENAR). The new *Biometric Bulletin* in its first issue delivered to members necessary details, rather than the previous custom of publishing details in *Biometrics* (though the journal did provide some limited information in 1983).

An Opening Session on Monday morning featured messages from the Japanese Minister of State Mr T Fujinami and the Minister of Education Science and Culture Mr Y Mori, followed by an address from Society President Pierre Dagnelie (Dagnelie, 1984). Beginning on the Monday afternoon, scientific sessions were conducted in the morning and afternoons throughout the week except Wednesday, and on the Saturday morning. The invited sessions covered: (1) Nearest neighbour analysis in field and variety trials (organized by H Desmond Patterson, BR); (2) design and analysis of intercropping experiments (Janet Riley, BR); (3) morphometrics-statistical methods for analyzing size and shape (James E Mosimann, ENAR); (4) analysis of DNA sequence data (C Clark Cockerham, ENAR); (5) confounding and other problems in epidemiological research (William J Blot, ENAR); (6) statistical problems in air pollution (Shun-ichi Yamamoto, JR); (7) use of historical control data in laboratory and clinical studies (Barry H Margolin, ENAR); (8) recent developments in the theory of proportional hazards models (John D Kalbfleisch, ENAR); (9) analysis of spatial point processes (Peter J Diggle, AR); (10) estimation for moving populations (Amode R Sen, WNAR); (11) applications of cluster analysis and multidimensional scaling to biometric problems (Yves Escoufier, RF); (12) model selection and validation (Kei Takeuchi, JR); (13) analysis of residuals (Susan R Wilson, AR); (14) use of personal computers in biometric studies (Andrew F Siegel, WNAR). Contributed paper sessions were held in parallel sessions and could be as oral presentations or as poster presentations; there were 162 contributed papers spread over 38 sessions. In addition, there was a data processing exhibition displaying statistical software and computer hardware.

Accompanying guests (the previously called 'family' members) were offered a number of cultural events. All participants were treated to an opening reception party on the Sunday evening, an all-day 'scientific' tour on the Wednesday featuring a visit to Tsukuba Science City and Whisky Distillery, and a Banquet on the final night Friday. The early registration fee was 30,000 Yen ≡ US$275 at the time, and 35,000 Yen ≡ US$320 after January 31. There were 450 participants and guests. The Japanese Region declined the Society offer of a $5000 loan.

With the new requirement (from Council in Toulouse) that sites be selected at least four years in advance, during this IBC, the Long Range Planning Committee re-affirmed, in a formal recommendation to Council that the 1988 IBC be held in Namur Belgium. This Committee also appointed Gerald van Belle (Chair, and next IBC Local Chair), Chikio Hayashi (this 1984 IBC Local Chair) and someone from Belgium (1988 IBC site) to prepare a manual for organizing IBCs.

9.1.13 IBC 13 – Seattle WA USA 1986

The Thirteenth IBC was held on the campus of the University of Washington, in Seattle, Washington, USA, from July 27 to August 1, 1986, officially invited by WNAR and hosted by the University of Washington's School of Public Health. The Local Organizing Arrangements Chair was Gerald van Belle. Chair of the Scientific Program Committee was Richard Tomassone (RF), with John Crowley (WNAR) in charge of contributed papers. In promoting the Conference, the local organizers pronounced that "umbrellas are quite unnecessary" [!].

The structure of the IBC followed that for the 1984 IBC in Japan, with parallel invited and contributed sessions in the morning and afternoons of Tuesday, Thursday and Friday, as well as Monday afternoon. The Opening Session was on the Monday morning, highlighted by the Presidential Address from Society President Geoffrey H Freeman. There were fourteen invited sessions covering topics (1) graphical methods for multivariate data (organized by John M Chambers, ENAR); (2) human genetics and genetic epidemiology (Jon Stene, NR); (3) statistical methods for acid precipitation studies (John O Rawlings, ENAR); (4) analysis of residuals in agricultural trials (Jean-Jacques Claustriaux, RBe); (5) mathematical and statistical modeling of physiological systems (Claudio Cobelli, RItl); (6) application of cross-validation (Geoffrey J McLachlan (AR); (7) controversial statistical issues in fisheries, forestry, and wildlife (J Richard Alldredge, WNAR); (8) recent advances in nonlinear theory useful in biometry (Olaf Bunke, RGDR); (9) problems of multiplicity in clinical trial data (Stuart J Pocock, BR); (10) applications of point process in biology, agronomy and environmental studies (Emmanuel Jolivet, RF); (11) stereology (Luis-M Cruz-Orive, ROeS); (12) data base management in biostatistical analysis (Toshiro Tango, JR); (13) recent development in multivariate bioassay (John J Hubert, ENAR); and (14) survival analysis (Martin Schumacher, DR). Members were invited to make suggestions as to speakers in these sessions. Each invited session had three speakers. While all members could present a contributed paper, a limit of one per person was imposed; over 160 abstracts were submitted. Wednesday was set aside for a selection of day-long tours including one to the Boeing Airplane Company (based in Seattle). In addition, there were daily tours arranged, all at an additional fee.

For the first time, the registration form and abstract form were available to members as inserts in the *Biometric Bulletin* (August 1985 issue). The early registration fee was $125 with $150 for late registration; this fee did not include the cost of the Closing Banquet.

The profits from this IBC were substantial (about $30,000); the $6000 from the Society was refunded, with the rest divided between WNAR and the sponsoring Department of Biostatistics by prior arrangement. In a much appreciated and nice gesture, soon thereafter van Belle advised the Society that the Department share would go to the Society to enable Third-world members to attend future IBCs. These awards were handled through the Awards Fund Committee over subsequent IBCs.

Both WNAR and RBe offered to host this IBC (as early as 1982); WNAR's bid was accepted for 1986, and the Belgian Region deferred to 1988.

9.1.14 IBC 14 – Namur Belgium 1988

The Fourteenth IBC was held in Namur Belgium from July 18 to 22, 1988, hosted by the Société Adolphe Quetelet, i.e., the Belgian Region (RBe), with meetings at the Facultés Universitaires Notre-Dame de la Paix. The Local Organizing Committee Chair was Ernest Feytmans (Belgian Regional President). The appointment of Hanspeter Thöni (DR) as Scientific Program Chair along with committee members in February 1985, more than three years before the IBC itself, marked a distinct difference from earlier IBCs when the invited program was being developed and/or finalized but a few months prior to IBC; later, Klaus Hinkelmann (ENAR) picked up the Program Chair work when illness forced Thöni to step down. The Scientific Program's responsibility now was confined to the selection of the Invited Sessions and Invited Speakers, with topics solicited from members.

The Scientific Program was structured as in recent years. After some introductory welcomes, the Opening Session on the Monday morning featured the Presidential Address from Society President Jonas Ellenberg. This Address focused on the contribution of biometry to medicine, and in a very real way reflected the contemporary emergence of medical issues in biometry in contrast to the primarily agricultural related issues that dominated the early decades of the Society and its IBCs. The Invited Papers Sessions covered: (1) The analysis of series of field experiments: Development in the theory and practice (organized by Mike Talbot, BR); (2) implications of matching for the design and analysis of observational studies (Stephen D Walter, ENAR); (3) dynamical aspects of clinical trials (J Dik F Habbema, ANed); (4) statistical needs for developing countries (Shrikant I Bangdiwala, ENAR); (5) quantitative genetic analysis in evolution and natural populations (William G Hill, BR); (6) issues regarding analysis of environment monitoring data (Loveday L Conquest, WNAR); (7) advances in statistical methods for the analysis of DNA sequence data (Susan R Wilson, AR); (8) exploratory data analysis for multivariate methods (Yves Escoufier, RF); (9) analysis of epidemiological data (Eugène Schifflers, RBe); (10) design and analysis of longitudinal studies (John L Gill, ENAR); (11) probability models for animal abundance (Richard Routledge, WNAR); (12) variance component estimation (Robin Thompson, BR); (13) methods for handling overdispersion (Norman E Breslow, WNAR); (14) growth curve models and analysis (Paul K Ito, JR); with two speakers per session or occasionally a third usually as a discussant. These invited papers were published in a *Proceedings* distributed at the Conference. Contributed papers could be the now familiar oral or poster presentation, with a third option 'microcomputer' session in which members could illustrate their work dynamically or as software demonstrations. The Belgian National Day, July 21, Thursday, was set aside for excursions (to Bruxelles, Bruges, or Ardennes).

Again, as was now customary, there was an early registration fee (Belgium Francs BEF8000 ≡ US$210 in 1988) and a late registration fee. The Friday night Closing Banquet and excursions were extra. A new feature was a *Daily Bulletin* edited by Éric Le Boulengé who was the regional Secretary-Treasurer and also the regional *Biometric Bulletin* correspondent.

The Society provided $7000 and an interest-free loan ($7000) to the Organizing Committee and $16,000 for travel and subsistence in Namur for the Executive Committee.

9.1.15 IBC 15 – Budapest Hungary 1990

The Fifteenth IBC was held from July 2 to 6, 1990 in Budapest Hungary. The Conference itself met in the intriguingly named National Council of Agricultural Cooperatives Hotel (on the hill-Buda side of Budapest). Local Arrangements Chair was Bela Györffy, though Hungarian Regional President Elisabeth Baráth was also heavily involved. The Scientific Program Chair was Niels Keiding (NR, later President 1992–1993); Zsolt Harnos (HR) handled the Contributed Paper Sessions. Preliminary announcements were made at the Namur IBC, and thus members were invited to contribute input to the invited program and to make their plans to attend.

The Opening ceremony was held on the Pest side at the University of Economics, with welcomes from Istvan Lang, General Secretary of the Hungarian Academy of Sciences. The Society President Richard Tomassone gave his Presidential Address on "The future of biometry," asserting that software would never replace biometricians (remember this was 1990). In sending the delegates out to the core scientific sessions, Local Chair Györffy practised what he preached with his memorable dictum that Hungarian people like "short speeches and long sausages."

The Scientific Invited Papers sessions covered topics in (1) R A Fisher Centenary (organizer Anthony W F Edwards, BR); (2) statistical aspects of molecular biology and molecular genetics (Claude Chevalet, RF); (3) spatial sampling aspects of environmental pollution monitoring (J Richard Alldredge, WNAR); (4) modeling in ecology and genetics (Yu M Svirezhev, USSR); (5) statistics in plant breeding (Rob A Kempton, BR); (6) methods for handling selected samples: Application to agronomy and epidemiology (Daniel Commenges, RF); (7) analysis of repeated measurements (Adelchi Azzalini, RItl); (8) statistical workstations (Reinhold Haux, DR); (9) Markov process modeling and the use of covariates in survival analysis (Per Kragh Andersen, NR); (10) exploratory survival analysis (John J Crowley, WNAR); (11) image analysis (Chris Glasbey, BR); (12) causal inference (Nanny Wermuth, DR); (13) structural inference (Gerhard Arminger, DR); (14) modeling and analysis of metabolism and pharmacokinetics (Hugo Flühler, ROeS). There were three aspects of this program that distinguished themselves. One is that for the first time there is a session organized by a USSR member (albeit not the first Eastern European organizer). A second is the fact that a session on workstations (session (8)) appears, thus marking the transition from the variously named computer contributed sessions to being a mainstream topic, a real reflection of the future scientific world. The other significant event is the celebration of the centenary of Fisher's birth (session (1)), a landmark truly deserving in recognition of our Society's founding President. Contributed papers (in 39 sessions) could be oral, poster, or computer sessions.

An interesting feature of this IBC was that it was approved in 1988 when events surrounding the Fall of the Berlin Wall had still to unfold. Thus, at the time, there were concerns about adequate funding especially since registration numbers would likely be low, concerns which exercised the creative talents of the Society officers at the time. By the time the IBC occurred, the former regime in Hungary had gone, with a consequence that the borders were open to visitors and conference attendees so that the registered attendance far exceeded initial expectations. The final count recorded 540 registrants and 90 accompanying guests.

Early registration was set at US$185 for non-rouble-clearing countries. Social events cost extra. Hotel charges were in the range of $65–80 per night, and student hostel accommodation cost $8–12 per person night. There were several excursions arranged throughout the conference. There was also an after-conference tour to Western Hungary; those participants watched bemused as fellow travelers, one by one, arrived bleary-eyed for breakfast one morning expounding in awe on the marvellous (three-tenor) concert from Rome which concert had been televised late into the night.

The Society provided $5000 for local expenses including publishing the *Proceedings* as well as $800 for invited speakers (though there were some executives who were "not thrilled [by] the notion of expenses for invitees except for [those from] underdeveloped nations." Travel expenses were always an issue. Thus, with the Council approval of Budapest, then Council member Lynne Billard (ENAR, later Society President 1994–1995), prompted by President Ellenberg, set about the process of submitting successful grant applications to the NSF for support for members to attend this IBC. Billard continued garnering travel support (in later years, targeted primarily for women, young and traditionally under-represented investigators) from the NSF but also from the National Institutes of Health (NIH) for subsequent IBCs up to and including the 2006 IBC in Montreal Canada.

9.1.16 IBC 16 – Hamilton New Zealand 1992

The second IBC to occur in the Australasian Region (AR) was the Sixteenth IBC held at the University of Waikato in Hamilton New Zealand (NZ) from December 7–11, 1992. The Local Organizing Chair was Ken Jury with Harold Henderson as Conference Secretary. The Scientific Program Chair was Jean-Jacques Claustriaux (RBe), with contributed papers chaired by J A (Nye) John (AR).

Len Cook, the NZ Government Statistician, opened the Conference followed by President Niels Keiding Presidential Address on "The Biometric Society: Diversity and Unity." The Invited Scientific Program covered topics in: (1) design and analysis of large-scale field experiments (Session Chair Rob A Kempton, BR); (2) biometry in human genetics and plant genetics (Elizabeth A Thompson, WNAR); (3) extensions of generalized linear models (Annette J Dobson, AR); (4) statistical needs for developing countries (Norman A Goodchild, AR); (5) consulting and collaboration (Pierre Dagnelie, RBe); (6) Bayesian monitoring of clinical trials (Dennis O Dixon, ENAR); (7) interface of geographic information systems and statistical analysis tools (Peter J Diggle, BR); (8) the AIDS epidemic: past, present, and future (Lynne Billard, ENAR); (9) the use of computers to design experiments (Dieter Rasch, RGDR); and (10) statistics in ecology and environmental science (Isao Yoshimura, JR); each invited session had two invited speakers. There were also 263 oral and poster contributed sessions. Computer software and statistical packages had migrated to exhibition booths, along with books. *Proceedings* were distributed to registrants but also were available for purchase after the IBC. There were at least nine satellite meetings organized around this IBC, held in other parts of the Australasian Region (New Zealand and Australia).

On Tuesday, delegates travelled to the Turangawaewae Marae in nearby Ngaruawahia to experience some of New Zealand's indigenous culture. Accompanying persons had a range of social events arranged for their enjoyment. Some of the pre- and post-conference tours were for the brave, especially those for "thrill-seekers" featuring black, and white, water rafting; though other tours were more sedate such as a paddle-boat cruise to a winery and stud farms. Early registration fees were NZ \$300 ≡ US\$155 at the time. University residence halls provided accommodation at NZ\$40 ≡ US\$21 per night (including breakfast).

Throughout the week, the Executive Committee (shown in Figure 9.6), indeed all committees, met along with Council Sessions.

FIGURE 9.6
Executive committee: Richard Tomassone, Elisabeth Baráth, Girja K. Shukla, Steve George, Lynne Billard, Roger Mead, Niels Keiding, Charles McGilchrist, Klaus Hinkelmann, Elsie Thull (Business Office) – Hamilton NZ IBC 1992.

9.1.17 IBC 17 – Hamilton Canada 1994

From one Hamilton in the Southern Hemisphere in 1992, the Society moved to another Hamilton in the Northern Hemisphere for the Seventeenth IBC, this time to Hamilton Canada over August 8–12, 1994, hosted by ENAR, and held at the Hamilton Convention Center. The southern Hamilton was located near the hot mud pools of the Rotorua district, while in contrast, the northern Hamilton was located on the Niagara Escarpment and the nearly Niagara Falls – both perfect locations for keeping informed on biometry. The Local Organizing Chair was Peter D M Macdonald; Byron J T Morgan (BR, later President 1996–1997) was the Chair of the Scientific Program Committee and Stephen D Walter (ENAR) handled the contributed paper program.

Conference participants were welcomed on Monday morning by Macdonald. This was followed by Society President Lynne Billard's (ENAR) Presidential address "The roads travelled." An innovation was a second projector featuring Robert Frost's poem "The Road Not Taken" (hence, the title of the Address) which was displayed simultaneously with the presentation itself. [At this point in time, it is recalled that talks had progressed from chalk and board or slides to transparencies, with powerpoint capabilities still to make their entrance.] Instead of giving a brief overview of the state of the Society as often outlined in previous Addresses, this address travelled through the content of the first ten years of the journal *Biometrics*, and then projected as to where the future of the subject biometry might lead (see Billard, 1995). The Scientific Program Committee organized Invited Sessions which dealt with (1) total quality management systems in biometry (organizer J Gölles, ROeS); (2) 100 years of mixtures (Geoff J McLachlan, AR); (3) neural networks and genetic algorithms (John N R Jeffers, BR); (4) issues in repeated measurements/longitudinal data analysis (Emmanuel E Lesaffre, RBe); (5) issues in fisheries research: resource management/spatial models/climate change (Stephen J Smith, ENAR); (6) recent advances in calibration, as applied to bioassay (Emmanuel Jolivet, RF); (7) GLMs and GAMS in biostatistics (Trevor J Hastie, WNAR); (8) estimation of HIV infected incidence, and projections of AIDS incidence (Patty Solomon, AR); (9) DNA and protein sequence analysis (David J Balding, BR); (10) computer-intensive techniques: Gibbs sampling/resampling (Amy Racine-Poon, ROeS); (11) spatial/temporal models in environmental statistics (Alfred Stein, ANed); (12) multivariate survival data: use of frailty models/direct modelling of marginal distributions (Steven G Self, ENAR); (13) stochastic and deterministic modeling in development, cancer, immunology and ageing (Tom B L Kirkward, BR); and (14) impact analysis for ecological and environmental time series, with applications to tropical rain forests (Sagary Nokoe, GKe). These invited sessions usually had two invited speakers with a third invitee as a discussant in some. Outside of this invited program structure was another presentation invited by the local organizers. Contributed paper sessions were of the now standard format as either oral (fifteen-minute talks) or poster sessions. However, an innovation was the special contributed session, consisting of at most six fifteen-minute contributed talks organized round a particular topic. Another innovation consisted of graduate student sessions.

Early registration fees were Canadian \$300 ≡ US\$155 at the time. There were about 546 registrants from 46 countries. Early budget/financial statements and reports declared that the Conference profits would allow for the Society loan (\$8000), as well as the loan from ENAR (\$2000), to be re-paid, along with the nett proceeds. However, despite local organizer's consistent promises that persisted for more than two years that the funds were coming, in the end there is a suggestion in the archives that creative financing subsequently erased them. Thereupon, as so often in the past when perceived problems arose, the Society set about revising its procedures to ensure the Society's interests (and by implication, its members) were foremost. In particular, more rigorous controls were developed and partnership arrangements were established between the Society and host-regions in regard to profit sharing paradigms for future IBCs. Also, not unsurprisingly, the Finance Committee in 1996 emphasized the need for the Society to take full responsibility for future IBCs and to be wary about unbudgeted and unapproved (albeit with some merit) expenditures creeping in. The expectation that each IBC be self-sustaining financially was now an established fact.

9.1.18 IBC 18 – Amsterdam The Netherlands 1996

The Eighteenth IBC was held at the Free University in Amsterdam The Netherlands from July 1 to 5, 1996. The Local Organizing Committee Chair was Arend Heyting with Hans C van Houweligan (Regional President) in charge of the Contributed paper program, and the Scientific Program Committee Chair was Susan S Ellenberg (ENAR). Since the journal *Biometrics* started fifty-one years earlier in 1945 and the Society started forty-nine years earlier in 1947, this IBC served as a special Society fiftieth Anniversary Celebration.

As was now the custom for the Monday morning, the Conference was opened with a welcome by T Sminia Dean of the Medical Faculty of Free University, followed by van Houwelingen on "Biometry in the Netherlands." Then, Society President Byron J T Morgan (BR) gave his Presidential Address. The Invited Program featured fifteen sessions: (1) AIDS in the 1990s (organized by Ron Brookmeyer, ENAR); (2) radiation risks: a review (David R Cox, BR); (3) dynamic graphics: a change of attitude in data analysis (Anna Bartkowiak, GPol); (4) statistics for environmental data (Carles Cuadras, REsp); (5) stochastic multivariate models of population dynamics (James H Matis, ENAR); (6) allowing for competition effects in the design and analysis of experiments in agriculture, horticulture and forestry (Rosemary A Bailey, BR); (7) Bayesian methods in biometry (Christian Robert, RF); (8) approaches for extending generalized linear models with random effects and corresponding components of variance (Bas Engel and Bertus Keen, ANed); (9) nonlinear models and optimal design (Timothy O'Brien, ENAR); (10) image analysis in biometry (Mats Rudemo, NR); (11) noncompliance in clinical trials: causal effect of observed exposure on outcome (Els Goetghebeur, RBe); (12) issues in genetic epidemiology (Robert Elston, ENAR); (13) design and analysis of ecologic studies (Ross L Prentice, WNAR); (14) exposure measurement error in nutritional epidemiology (Raymond J Carroll, ENAR), and (15, see below) historical plenary session in celebration of the fiftieth anniversary of the IBS. Sessions (1)–(14) were held from Monday morning through Friday mornings, in ninety-minute blocks featuring a total of 35 invited speakers and nine invited discussants. Forty-six Contributed Paper Sessions, with 227 papers, were held throughout the week in parallel with the Invited Sessions, and there were 108 poster papers displayed across three sessions. The Conference ended with lunch on Friday. The "computer" related poster formats of previous conferences had undergone yet another migration, this time to Commercial Presentations, one for "Genstat for Windows" and another for "Stata" for biometric analysis and "Gauss for Windows." Two one-day short courses were scheduled, one to precede and one to follow the Conference.

The Thursday afternoon and evening focused on celebrating the fiftieth Anniversary of the Society. It began with an Historical Plenary Session (session (15), organized by the Program Committee Chair, Susan Ellenberg) which as the name suggests was designed to highlight the first fifty years of the Society. All living current and former presidents were especially invited to attend. These presidents served as Discussants to a historical presentation "The Biometric Society – fifty years on," by Peter Armitage, himself a former President and former *Biometrics* Editor; see Armitage (1996). The Anniversary Celebrations culminated with an after-banquet slide presentation showing photos and other memorabilia of the Society. During the three to four years prior to this event, Billard (as Vice-President and/or President) had taken photos of photos as she visited the various regions and national groups (at no cost to the Society). Requests for these photos had been made; so the regions were ready when she actually arrived. Former President Jonas Ellenberg, dressed in tuxedo and bow-tie, was the session's Master of Ceremonies. A good time was had by all!

As a first, email and computer facilities were available throughout the conference. Exhibits, mostly from computer and/or software venders and book publishers, were held the last two days.

All-day conference tours were held on the Wednesday, including a visit to the Zuiderzee Museum Enkhuizen and Zaanse Schans. A variety of other tours were arranged for accompanying persons. Except for the Opening Reception (on Sunday night), all social events were at an additional costs (again as had become customary); this included the Anniversary Banquet (Dutch Guilders 100 ≡∼US$60 in 1996). The early registration fee for members was Dutch guilders Dfl375 ≡∼US$220, Dfl450 for non-members and Dfl60 for students. Attendees numbered 672.

With this IBC, the Outgoing Vice-President (i.e., immediate Past President) was the Organizing President for the IBC, the first to serve so was Billard (President 1994–1995).

9.2 Non-IBC Symposia

9.2.1 Indian Symposium, 1951

At its January 1950 regional meeting in Poona, the Indian Region (RInd) resolved to invite the Society to hold its biennial conference in India in 1951. This two-day Symposium "Biometric Problems in the Prediction and Estimation of the Growth of Plants in Tropical and Subtropical Regions" was held in Calcutta and timed to coordinate with the ISI Session scheduled for December 1951 in New Delhi and in Calcutta. Because of UNESCO/IUBS funding constraints on designating this an IBC (in this case because it was less than three years since the previous IBC), this would have to be a Symposium which was supported by $1200 UNESCO funds. Society President Arthur Linder was on leave at the Indian Statistical Institute July to December

1951 and was in charge of the scientific program. Assistance in identifying specific topics of interest to India was to come from PC Mahalanobis. The ISI Director, Gysbert Goudswaard, invited the Society to participate in its ISI session. As late as June 1951, both Goudswaard and Bliss were concerned over the "indefiniteness of the commitments" from Mahalanobis (elsewhere the Archives reveal "the clerical work tends to be unreliability" with associated delays being frequent events). In June 1951, the Society sent a lot of letters inviting specific folk to participate – distinguished lists indeed – with considerable interested responses. Proceedings of the Seminar were sponsored by the United Nations and the World Health Organization. Potential participants are advised that transportation by P. and O. steamer from London to Bombay would require sixteen days, but that flying time from New York to Delhi "[was] just three days."

On the first day, the 150 participants enjoyed scientific talks which covered the history of crop prediction (J Oscar Irwin), crop prediction in England (Frank Yates), and sampling experiments of Chinchona in Madras (Sengupta, Chakravarti, Sarhan). The 100 participants of the second day heard reports on recent experiments in crops in Australia (Belz), and crop-cutting experiments (Mahalanobis). Each day provided for discussions, with contributions made by Banerjee, Haldane, Fisher, Georg Rasch (who claimed he was not much of an expert here but he "was quite prepared to listen and look wise"), Sukhartne, Yates, CR Rao, NK Rao, Kishen, and Chiney.

Several communications between the Society (mainly Bliss) and India dealt with Bliss' hopes that the Symposium would help reactivate interest of the local members in the original Indian Region formation which region was now in the throes of dissolution over dues (see Chapter 5, Section 5.1). While the Symposium was a success, the regional reactivation hopes were not.

9.2.2 Brazil Symposium, Campinas 1955

The University of São Paulo hosted the "International Biometric Symposium in Brazil" held in Campinas Brazil July 4–8, 1955. This followed a meeting of the ISI in Rio de Janeiro (June 24 to July 2, 1955), and preceded a Inter-American Statistical Institute (IASI) Conference in Santiago (July 7 – 23, 1955) at which William Cochran presented his twelve-page report on "The Teaching of Statistics for Application in Biometrics." This Symposium was not in fact an IBC much as Bliss and his colleagues would have liked, but instead the gathering was called a "Symposium" in order to satisfy a requirement associated with securing funding support from IUBS.

On Monday morning July 4, symposium participants were welcomed by R Cruz Martins (Secretary of Agriculture for the state of São Paulo), followed by the Society's Presidential Address William G Cochran on the poliomyelitis trial of 1954, which paper became a landmark paper on clinical trials research and to *Biometrics*, see Cochran (1955). The formal scientific sessions began in the afternoon. Throughout the week, topics and speakers

FIGURE 9.7
David Finney presenting at Campinas, 1955.

covered (1) biometrical genetics (Ronald A Fisher (BR), Everett R Dempster (WNAR), Friedrich Gustav Brieger (Brazil), and Hans Kalmus (BR)); (2) experimental designs for perennial crops (S Clifford Pearce (BR), Edilberto Amaral (Brazil), Armando Conagin and Constantino G Fraga (Brazil), and F. Pimentel Gomes (Brazil)); (3) experimental design (Gertrude M Cox (ENAR), and W Jack Youden (ENAR)); (4) panel discussion on experimental designs for perennial crops; (5) statistics applied to animal feeding experiments (Paul G Homeyer (ENAR), Geraldo Leme da Rocha (Brazil), G L Mott (ENAR), and Arthur Linder (ROeS)); (6) sampling techniques (Morris H Hansen (ENAR), Panduring Vasudeo Sukhatme (Food and Agricultural Organization (FAO) of the United Nations), José Nieto de Pascual (Mexico), Wilfred Leslie Stevens and S Schattan (Brazil), and Enrique Cansado (Chile)); (7) bioassay (Chester I Bliss (ENAR), David J Finney (BR), and Paul Mello Freire (Brazil)); (8) medical statistics (J Oscar Irwin (BR), Jacques Noel Manceau (Brazil), A E Brandt (ENAR), and André Vessereau (RF)). Figure 9.7 which shows Finney delivering his presentation, reminds us that talks were board and chalk deliveries. The talks on perennial crops featured sugarcane and coffee; Stevens also spoke of sampling of coffee harvests. In an interesting innovation, the Wednesday schedule took participants on an excursion to Piracicaba; here the panel discussion (session (4) above) took place in the morning at the Luiz de Queiroz School of Agriculture, while in the afternoon, they visited a sugar mill. On Friday morning, they visited the State Institute of Agronomy at Campinas. The excursions to the sugar mill, and to a dairy and coffee farm (on Saturday; see Figure 9.8) were conference highlights. Threaded through the schedule were meetings to discuss the formation of a Brazilian Region of the Society (see Chapter 5, Section 5.6).

The archives contain endless files dealing with this Symposium. These began with handwritten notes from the 1953 Rome ISI meeting wherein Bliss,

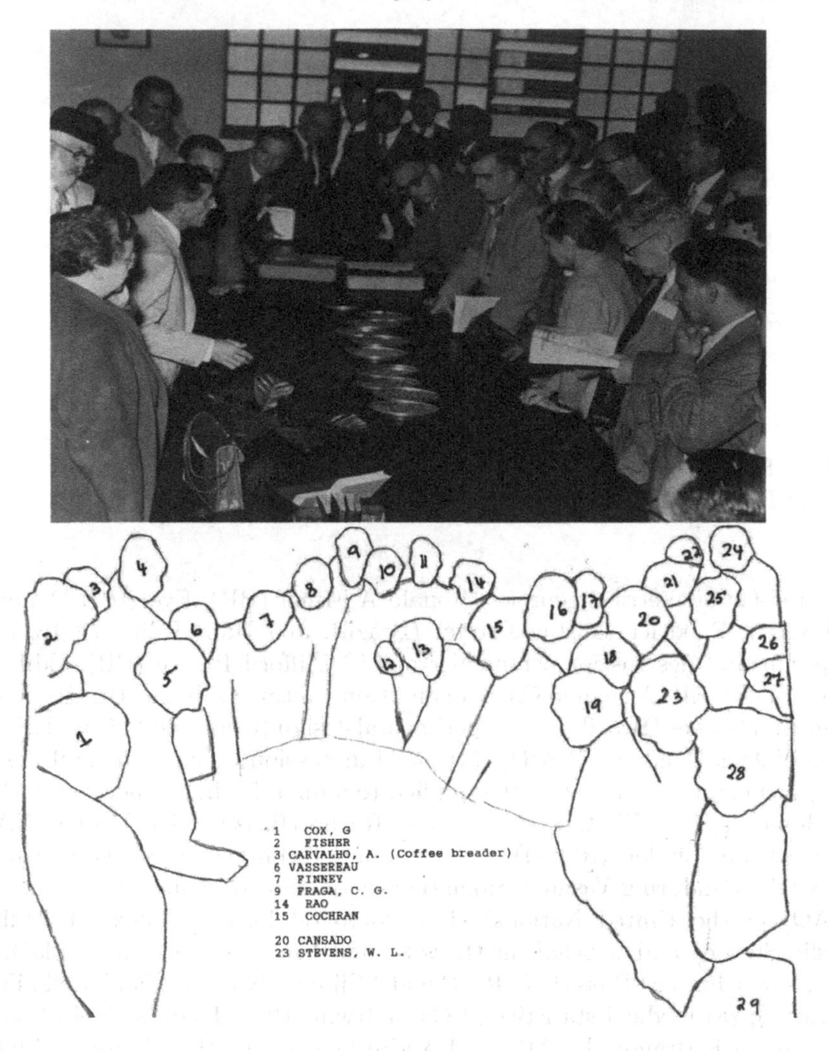

FIGURE 9.8
Participants and identification on coffee excursion, 1955.

Linder, and Fisher outline some possible details including scientific session topics and speakers. A tentative outline of such a Symposium had been proposed to IUBS at its Nice August 1953 meeting. There is a slew of letters (mostly from Secretary Bliss) enquiring as to who might serve as official host, and also where the meeting could be held. Progress toward convergence was slow, so much so that in July 1954, following a suggestion from Gertrude Cox, Jack A Rigney was approached by Cochran to visit and speak with various folk during his upcoming visit to Brazil to solicit input. Rigney's August 1954 report to Cochran was superb, bursting with valuable information (Campinas, coffee and sugar cane and perennial crops as well as breeding and selection techniques, excursions, internal expenses funding, and so forth, much of which was adopted) that allowed the Symposium to move forward. [This Rigney was based in Raleigh, very much unacclaimed, seemingly shunned the limelight, but he was clearly instrumental in bringing things into focus – his role was absolutely key.] Ultimately, rather than in Rio de Janeiro, it was determined that most members were in the São Paulo area, so that it was the state of São Paulo and its University selected as host and site. The national representative to the Society, Américo Groszman, participated in discussions early on; in the end it was Constantino Gonçalves Fraga Jr (first Brazilian Regional President) who effectively served as the Local Organizing Secretary as he was himself based at Campinas.

Cochran was actively engaged in the symposium and its scientific content, in particular responding favorably to those who wanted the focus to be on practical issues rather than theoretical questions. One question which arose early was when Henry Hopp (a United States Department of Agriculture (USDA) employee) in November 1954 echoed the suggestions voiced by Rigney about the problems of coffee growers, such as how many trees should be planted per unit land-mass (opinions varied from 4 to 100), inherent variance component issues, all part of a long useful detailed list; this was clearly a topic of primary importance to an important coffee growing nation. Cochran quickly assured Hopp that they would "bend their energies" to address such concerns. Meanwhile, back in Campinas, there was even a donation of 60 coffee bags (worth about US$3000 at the time) so a substantial contribution which helped with local expenses. Not too unsurprisingly, there was considerable interest in an excursion to a coffee establishment during the Symposium itself. This was clearly a Symposium that addressed itself to the scientific concerns and needs of the host country. The Symposium was a huge success, with 96 participants from 17 countries; see Figure 9.9.

Ten government and nongovernmental organizations gave financial assistance; also the IUBS gave $2000 for travel as well as $500 towards publishing the proceedings (which came out in *Biometrics* in December 1955 and March 1956). Non-Brazilian participants received travel assistance from

FIGURE 9.9
Participants at Campinas symposium 1955.

a variety of entities such as the Royal Society, British Council, and NSF. The ISI ran a special plane from Paris to Rio de Janeiro at the low fare of $150 for ISI members so many Society participants could avail themselves of this opportunity. Getting to Campinas was another story. The (May 1955) Archives tell us "... the plane fare from Rio to San Paulo is $24.90 (three planes daily). The trip can be made also by a rather hazardous bus at about $15 or less ... . The bus drivers do not believe in slowing down for curves, the road is mountainous, and they have a substantial record of fatal accidents." Anyone who has flown over that route can attest to the mountainous terrain below. The Archives gave no further details on this.

There were many communications regarding possible travel support for potential participants. Even Cochran, though President, was not assured of funding; he told one aspiring attendee that "I hope to go if a fairy godmother appears." She must have shown up since Cochran attended.

9.2.3 Other Symposia

The Varenna Seminar in Biometry was held September 7–23, 1955 at the isolated and picturesque Villa Monastero at Varenna on Lake Como Italy. The Italian Region (RItl), primarily Luca Cavalli-Sforza, was eager to meet the growing demand for instruction in matters biometry. The idea to run a short course was quite novel at the time, at least in Europe. Two principles drove the organizing process: one to conduct basic courses, if possible, in Italian; and the other to have an international teaching body within this restriction. Roughly four times the number of planned-for attendees applied; subsequently, the number accepted was doubled with students working in pairs at calculators for the afternoon practical sessions, giving a total of fifty-six students all but one from Italy. There were four core courses: (1) theoretical foundations (Maria Pia Geppert, DR); (2) applied statistical models (Cedric A B Smith, BR); (3) design of sampling surveys and experiments (Frank J Anscombe, BR); and (4) single degrees of freedom (Luigi L Cavalli-Sforza, RItl). The afternoons saw the students divided into two groups, one for students from agriculture and one from medicine. There were also lectures by Renzo E Scossiroli and P Dassat, P V Sukhatme, Arthur Linder and H Furgąc, A Tizzano, Gustavo Barbensi and Linder, and Giulio Alfredo Maccacaro. A restricted group received an additional short course of six lectures on "The logic of inductive inference" by Ronald A Fisher (BR). The IUBS (it is recalled that the Society was its Section of Biometry) was keenly interested in supporting educational pursuits, and so provided funds ($1000) to bring this to fruition. These funds, together with the student fee of $15 and $500 from the Italian Region, amply covered expenses. See Cavalli-Sforza (1956) for an excellent detailed report. A second course was held in Milan in 1956.

Inspired by Varenna, Arthur Linder (then National Secretary of the Swiss Group, and a former President) organized an Internationales Biometrisches Seminar and Symposium (Seminar and Symposium in Biometry) for German speaking people, held September 24 to October 3, 1956, in Linz Austria. Linder found excellent support from Adolf Adam who was a member of the University of Vienna, but more crucially was a statistical consultant with Öesterreichische Stickstoffwerke A G at Linz. This firm provided financial support, along with other support from government authorities; Adam made the local arrangements while Linder took care of the scientific program details. There were newspaper reports in the Linz newspapers, and some members including Linder were interviewed by Linz radio. Seminar courses covered (1) statistical methods (Adolf Adam, Johann Pfanzagl, and Erna Weber); (2) experimental design (E A G Knowles and Arthur Linder); (3) sample surveys (Hans Kellerer); these lectures were complemented by case histories of statistical investigations. A simultaneous Symposium during the last two days featured discussions on papers by Hans Linser, Gustav Herdan, and Walter-Ulrich Behrens. Over 150 people attended.

There was a "Symposium on Quantitative Methods in Pharmacology" in Leyden The Netherlands, on May 10–13, 1960, with partial support from ICSU/IUBS (US$1500). Sessions on sequential methods (chaired by Peter Armitage, BR), standardization of drugs (Cliff W Emmens, AR), parametric or nonparametric statistical methods (J Hemelrijk, ANed), screening of drugs (Owen L Davies, BR), mixtures of drugs (Heinz Otto Schild, BR), and miscellaneous topics (David Karel de Jongh, ANed). The Symposium was organized by a committee chaired by de Jongh (the Region's first president), and attended by 156 scientists. The proceedings were published as *Quantitative Methods in Pharmacology* by North-Holland Publishers. Bliss being Bliss thought staging this Symposium would help to foster interest in establishing a new region for the Society.

A Biometric Colloquium was held in Prague Czechoslovakia on April 8–10, 1965, sponsored and organized by the German Region. The Society endorsed this meeting, most especially as it was seen as an important avenue to making contacts useful for the future growth of the Society in Eastern Europe, and was "astonished" at the participation of 260 persons. The only down-side reported was the observation that no USSR person attended.

9.3 Concluding Comment

When talking of the 1949 Geneva IBC, Helen Turner (AR member) opined that "fraternisation is the chief aspect of conferences." To this is added the Report of the 1965 Prague Czechoslovakia Colloquium declaration that "The idea of the necessity of universal friendship, teamwork and exchange of literature was more evident every day." These sentiments surely sum up the key principles behind the importance of holding international conferences.

10

Governance, Constitution, and Management

The previous chapters have told of a vibrant Society, with active regions and national groups, and with important scientific undertakings through its publications and meetings. Behind any successful organization lies a governance and administrative structure to assist the members in their scientific endeavors. In this chapter, we describe governance through its constitution and By-laws (in Section 10.1) and the role of management and its changes over the decades (in Section 10.2).

10.1 Governance: Constitution and By-Laws

The constitution by definition provides the foundational principles governing the Society; By-laws provide rules and regulations that allow for the implementation of these constitutional principles.

Participants at Woods Hole (in 1947) debated article by article, line by line, word by word, the draft Constitution prepared by the Committee on International Organization; this discussion was outlined in Chapter 2. Changes were made over the years – sometimes to streamline what had become accepted practice bringing up to date processes that had been approved bit by bit in prior years by Council, sometimes to bring clarity to interpretation, and sometimes as a major overhaul. Progress was invariably slow for these revision committees since most communications were by postal mail, and further recommendations then had to be approved by Council before going to the membership; so, it was typically three-four years before completion.

Revisions in 1956 were led by Treasurer Allyn Kimball, Secretary Michael Healy, and Editor John Hopkins; Chester Bliss took responsibility for the 1962 revisions but with assistance from Treasurer Marvin Kastenbaum and Healy. Both these revisions essentially modified the Constitution and/or By-laws to incorporate aspects that had already been approved by Council since 1947 and 1956, respectively, and to straighten out what had become a "legalistic tangle." A key addition in 1956 was the formal structure of the already existing Editorial Board. The 1963 revision went to members for ratification in December 1962, and became effective January 1963.

DOI: 10.1201/9781003285366-10

Later, in 1979, President John Nelder appointed an ad hoc committee (with incoming President Richard Cormack as Chair and members Herb David, Lyle Calvin, Jonas Ellenberg, and Peter Armitage) to study and make recommendations for a revised Constitution and By-laws, which revisions were approved by members in 1983. Perhaps the most notable change here was that balloting for Council members by mail was added. In 1994, President Lynne Billard appointed a revision committee whose recommendations produced the 1998 Constitution (the last official Society Constitution). There were two other, quite major efforts, viz., when President Berthold Schneider appointed a Committee which produced the 1974 constitution, and when the 2009 initiative of President Andrew Mead led to the new modus operandi government structure of the 2012 revisions (described briefly in Chapter 11, Section 11.4). We first consider issues that came up regularly, followed by a look at those issues which arose from specific revision committees.

The first major item discussed at Woods Hole was the name of the society, in particular whether or not the word "international" should be included. This word in the name of the Society continued to receive attention. Many wanted to include it, and furthermore, the word was often used as the upper case "I" or more often as the lower case "i" adjective. Officers frequently suggested the change. Indeed, even in Bliss' August 1, 1962 letter to the Editor of *The Times* on Sir Ronald Fisher's death, the word was used. Judging by these frequent usages and complaints (such as versions of "we are an *international* society"), eventually, it came up for discussion at the Cambridge Council meeting in 1963. A split vote opted to continue as it was for the English, but the German "Internationale Biometrische Gesellschaft" and the French "Société Internationale de Biométrie" usages could be retained. The complaints continued. The English version became "International Biometric Society" with Council approval in 1994, to be consistent with the French and German terminology.

Society officers formed the Executive Committee (initially consisting of the President, Secretary, and Treasurer); they, as well as the Vice-Presidents, were members of Council. The make-up of that Committee has however changed over the years. The Editor (of *Biometrics*) was added with the 1956 revisions. Also, in 1956, these Executive Committee members became ex-officio members of Council. Prior to 1953, the Vice-Presidents were the elected leaders of the respective regions, in effect the regional president. In 1953, this Vice-President title was formally changed to Regional President but these positions were still ex officio to Council. After the 1974 revisions of the Constitution, the regional presidents were no longer ex officio to Council (though regional presidents could be an otherwise elected Council member). In the 1963 revisions, a Society (Incoming) Vice-President was elected (to a one-year term) and that person would automatically serve as President in the following two years, and then return to being the (Outgoing) Vice-President during the fourth year. The Society Vice-President was a member of the Executive Committee and hence ex officio to Council. The President was, still is, Chair of Council.

The roles of Secretary and Treasurer also saw frequent debate. At Woods Hole, one-year terms were adopted with no limit on the number of re-elections. This was still so with the 1956 revisions though now the provision that they could be combined was added. In 1963, the term became two years with re-election still possible; no limit was given for the number of times re-elected. Finally, in the 1974 revisions, each term was set at four years and "may be re-elected once." The option that these two offices could be combined was eliminated with the 1983 revisions; also in 1983, the election cycle for these positions was set so that the overlap was two years.

With the 1956 revisions, the Editor became a part of the Executive Committee. A new By-law in 1956 decreed that there be an Editorial Board (of twelve maximum members, each as a regional representative) to formulate general editorial policy, with the Editor appointing Associate Editors (who would be ex officio to the Board but could not constitute a majority of the Board). The 1983 revisions saw the name Editorial Board changed to "Editorial Advisory Committee" (EAC) and specified that the EAC advise the President on editorial appointments. With the emergence of new publications, the 1994 Constitution Committee debated whether to include all Editors or an Editorial representative; it was decided to include an Editorial Representative. While this inclusion was approved in 1998, by the time of the 2012 revisions, the position was no longer part of the Executive.

In 1970, President Schneider appointed a Constitution Revision Committee (Treasurer Henry Tucker (Chair, WNAR), Richard (Dick) L Anderson (ENAR), David J Finney (BR)). [This is the same Anderson who as ENAR president had disputed Council election procedures and who had subsequently formed an ENAR committee to recommend constitutional revisions; see Chapter 4, Section 4.1.] On the eve of writing the Committee's Report, Tucker[1] had a "fatal heart attack" (September 1971), whereupon Anderson became Chair and wrote the draft report (October 1971) from Tucker's notes. Then, the Executive Committee debated the recommendations along with Anderson and Finney and thence Council. During these discussions, Vice-President C R Rao writes in January 1973 that he would leave the conclusions to the rest of the Executive, and also in January 1973, Editor Frank Graybill having spent an extended time in hospital deferred to the rest of the Executive. However, President Armitage, Secretary Hanspeter Thöni, Larry Nelson, Schneider, Anderson and Finney kept the files busy with lots of exchanges. The most controversial issue was a proposed merger of the Secretary and Treasurer positions, ultimately dropped on the grounds that it was better to spread offices across more rather than fewer members. An early proposal to eliminate the Finance Committee was also quickly dropped as treasurers were in favor of its retention.

[1] Cady (1972) has a pithy but delightful in memoriam, telling us that "talking was one of his talents."

The major change dealt with the nomination for and election of Council members; this received a lot of attention. Finney (supported by others) was particularly concerned that elected members have the interests of the Society foremost rather than the interests of the region in which they resided. In the end, with respect to Council, it was recommended that Regions have a minimum of two members (to compensate for the loss of ex-officio status of regional presidents, which evolved from a 1963 Council decision at the Cambridge International Biometric Conference (IBC)), that the total number of ordinary members be three times the number of regions, and that terms be increased from three to four years (for more stability and allowing for the maximum four years between IBCs) with immediate re-election limited to two successive terms, and that no ordinary member could be concurrently an ex-officio member; with this new arrangement, half would rotate out every two years instead of the previous third every year. By implication, this meant that Council elections would now be held every second year. If, after an election round, a region was found to be without its full complement of representation on Council, the President would make appointments from that region. Other recommendations dealt with succession of office of the Vice-Presidency should the occupant be unable to continue (succession criteria already existed for President), the nominating procedure for executive officers was streamlined, and that new members to the Society only needed to be approved by their region (instead of Council). After approval by Council, the proposed changes were sent to members for ratification on May 1, 1974. The nomination process of Council (which process had so exercised ENAR leadership in the mid-1960s; see Chapter 4, Section 4.2.1) was shifted to the By-laws. In particular, regional presidents and national secretaries had the responsibility to submit nominees to the Secretary for each election cycle. The files suggested that not all regions or national groups were fully responsive or pro-active on this.

Much had happened since Cormack's work on the 1983 Constitution, not least that new journals had been added (which raised questions as to which editor/s should serve on the Executive Committee), and electronic communications promised to make substantial changes as to how a society would operate. Further, a need for revision was also largely prompted by the fact Roger Mead as Secretary in 1992 had encountered proposed changes to a particular region's By-laws that would contradict the Society Constitution. Revisions to protect against such a situation were sent to Council, clarifying that regional By-laws were a "supplement to the Society's Constitution" and concurrently that particular region's proposal to eliminate the two-year limit to the term for regional presidents was defeated (not surprisingly, as one President said, given the historical "opposition to unlimited re-elections of regional presidents"). Therefore, in an attempt to bring clarity and up-to-datedness to these facets, President Billard set up a 1994 Constitution Committee (Anthony W F Edwards (Chair, UK), Secretary Elisabeth Baráth (HR), Pierre Dagnelie (RBe), Jonas Ellenberg (ENAR), Peter T (Tony) Lachenbruch (ENAR), Charles A McGilchrist (AR), Hanspeter Thöni (DR)).

When the Society became incorporated[2] in 1996 (bringing with it Articles of Incorporation), the Constitution became the second level of rules governing the Society. This also meant integrating some elements of the Articles of Incorporation into the formal Constitution. These along with other recommendations from the 1994 Constitution Committee were approved in 1998. Changes to the Articles of Incorporation (first level) or to the Constitution required a vote of all members, while changes to the By-laws (third level) only required Council approval.

While the role of Council remained basically steady over the years, the number of ordinary Council members saw considerable changes, primarily because of the growth in the number of regions. Executive Committee officers were always ex-officio members of Council. The first Council numbered twelve but quickly changed to twenty at its second meeting (see Chapter 2). The 1974 revised Constitution specified at least two members per region with one additional 'region' representing members outside an organized region, giving a total of twenty-six members. This 'one' region for members not in an organized region became two 'regions' with the 1998 approved revision, and the actual number of Council members was set to be "no less than 15 nor more than 80." The upper limit was designed to cover the growth of new regions; there were eighteenth regions at the time. At this point in time, regions consisted of at least fifty members and national groups of at least ten members. Ten-twelve years later, national groups were to be called regions; this would have a real impact on the structure of Council. These changes and their impact are described in Chapter 11, Section 11.4.

10.2 Management of the Society

The first "executive assistant", i.e., "secretary" to Secretary Bliss was Eleanor Watkins who set up a table in her living room; she worked half-time and with the status of a Yale employee at $120 per month. Initially funds came from Yale University until a Rockefeller Foundation grant was received. When her husband (John H Watkins, ENAR Secretary-Treasurer) died suddenly fifteenth months later,[3] new arrangements had to be found. Subsequently, a new assistant started in December 1948 at Yale University in a temporary office in the Department of Public Health. In a letter to Arthur Linder in August 1955, Bliss reports that when this room was reclaimed, "the University

[2]Incorporation had been raised at the outset by Bliss, but tabled. Instead, meantime, the Constitution was amended to read that "No officer [or] member ... could receive any compensation or any pecuniary profit whatsoever ..."; this was necessary to ensure tax-exemption purposes. It was raised again by the Executive Committee in 1963 to protect members from possible legal suits, again tabled.

[3]One colorful letter refers to a plan formulated by Watkins which "Jack uncooperatively ditched by dying prematurely."

administrators assured me [Bliss] that there were absolutely no free rooms anywhere ... but I [Bliss] phoned one Department after another and after some eight or ten calls located our present quarters, much to everyone's amazement." Successive Society Secretaries had to make their own local arrangements for secretarial assistance; this made the search for successor Secretaries difficult.

As early as June 1964, Bliss talked with Secretary Henri Le Roy about the need to re-evaluate the role and workload of the Secretary position and that instead of the local part-time secretary, maybe it is time to move to a full-time (or part-time still, or semi-permanent) secretary in rented space. Complaints persisted through to 1968, precipitating an urgency for decisive action. An overture from the ISI to have a joint business office was turned down. Gertrude Cox advised the Managing Editor of *Biometrics* (Malcolm Turner) in mid-1968, that this new arrangement would likely happen, and then again in November 1968 that it would in fact happen with implementation in early 1969. Thus, the new structure as a business office began officially in 1969 located in Raleigh (with Larry Nelson as the Business Manager – a paid position, who could be an elected Council member but not an officer); the Business Manager's role was to take over the work of the Managing Editor and the routine work of the Secretary and Treasurer. For the first time, the Business Manager was bonded, which required a second signatory on bank checks, a role fulfilled by the business office assistant. Later that year, when Hanspeter Thöni came on board as Secretary, Cox explained how the business office should help in managing his workload, but added, ever mentoring, that there would be problems which she believed were "associated with growing pains"; she continued, saying that the Society had become a "rather large group," too large to be handled almost entirely on a volunteer basis. She concluded "it is never easy to try to operate an international society."

Cox with a reputation as an "excellent administrator and organizer" had been key in reorganizing the business arrangements. Also, upon becoming President, she immediately asked officers to provide detailed documents outlining their duties.

Nelson wanted to step down as a Business Manager in 1978. In December 1978, Nelson's secretarial assistant Elsie Thull stayed on as Business Manager but moved to the new business office location now housed in ASA's building. This move also involved the return of the management of ENAR and WNAR to the Society business office from ASA which had been managing these two regions since 1965.

Concerns arose again in the late 1980s about upgrading the business office. Although IBM cards (1963) and computerized lists (1974) for the Directory had made early entrances, computers and email as part of the information highway were making their presence felt, and clearly were not going away. Thus, after Council discussion at the Hamilton New Zealand IBC, President Niels Keiding in December 1992 set up an ad hoc Committee on Business Management (with Treasurer Stephen George as Chair, and Vice-

President Lynne Billard and ENAR President David Demets as committee members); Keiding separately wrote to Thull advising her of this review. The Committee's charge was minimally to review the functions and responsibilities of the business office, study possible alternatives and to consider financial implications of expanding or reducing office functions. Given the potential of a major paradigm shift in operations, it was crucial that the entire process be very thorough and fair to all. Therefore, each Committee member visited individually with Thull, as well as interviewing knowledgeable persons within the Society and in other comparable organizations. Thull frequently said she needed to be three differently skilled persons. In its October 1993 Report, the Committee concurred with Thull and concluded (from the Report to Council) that "one person could not possibly do all that is required, that the Office needs persons specialized in management, financing, computers, etc." It became clear that only two viable alternatives existed. One was to reorganize the one-person office to a three-four person office; the second option was to move to a professional association management firm. The first option was too expensive, while the second was more effective and less expensive. Subsequently, Billard and George visited Thull personally on December 2, 1993 to inform her of this decision and that the anticipated start date was expected to be in March 1994 (ultimately this was December 1994).

Given these recommendations, initiated by Keiding, Billard in January 1994 appointed an ad hoc Management Selection Committee (consisting of Mary Foulkes Chair (ENAR), Secretary Elisabeth Baráth, EAC Chair Robert Kuehl, and Council member Tom Louis as members, with Billard, George and Keiding ex officio). This Committee set up procedures to guide the process; six bids were selected for further consideration (including two based in Europe) and some were site-visited. In a detailed ten-page Report to Council and to Regional Presidents and National Group Secretaries for discussion at the Hamilton Canada IBC 1994 Council meeting, the Executive Committee summarized these steps thus far. Once Council had approved, a final selection was made. This led ultimately to moving business operations to a professional organization management firm, i.e., the Society now had an International Business Office staffed by personnel trained in professional association management. (A similar move by ENAR had already been made in 1992.) There would be economies of scale, provision of professional expertise to the Office, and a much-needed base for continuity and permanence in office operations, among other advantages. While the immediate cost of the selected firm was marginally more than the previous cost of operating the office, these costs were effectively recovered within six months after implementing changes for the production of *Biometrics*; see Chapter 7, Section 7.5. More importantly, however, the Society was now the beneficiary of a range of expertise impossible to attain in a one-person office.

11

Post-1997

Previous chapters have described different aspects of the Society's functions and roles by focussing on the first fifty years since its inception, 1947–1997. In this chapter, a very brief look at some key events since 1997 is presented. Thus, Section 11.1 looks at developments in the growth of new regions and national groups, as well as occasionally new initiatives of already established regions; Section 11.2 reports on changes in publications; and Section 11.3 outlines details of International Biometric Conferences (IBCs). A major change regarding the Constitution and governance occurred in 2012, described in Section 11.4. A final Section 11.5 covers other actions of note that have occurred since 1997.

11.1 Regions, Groups, and Networks

While both regions and national groups were part of the first fifty years, with the 2012 constitutional revisions, all national groups became regions; the number of members needed for formation became ten. Recent formations not included earlier are described in this section. These constitutional revisions also encouraged the formation of networks as loose arrangements of geographically close regions for the purposes of strengthening links between regions including sharing holding joint conferences, though many such arrangements were already in place as has been described in Chapter 6, Section 6.8, as part of the component national group formations. In addition, some larger regions formed loose network arrangements primarily for purposes of holding conferences and related events. We first look at regional and national group developments since 1997, then we look at networks that have evolved.

Chile formed a National Group (GCl) in 2000 with M G Icarza Noguera as its first National Secretary. The Ecuadorian Group (ECU) formed in 2015; Omar Honorio Ruíz Barzola was the first National Secretary.

An Eastern Mediterranean Region (EMR) was formed in 2001, comprising Cyprus, Egypt, Greece, Israel, Jordon, the Palestinian Authority, and Turkey. Saudi Arabia was added in 2003; Bulgaria was added in 2008. The first President was Laurence S Freedman. Every two years, this Region links up with the Spanish Region (REsp) and the Italian Region (RItl) in an informal network for conferences.

DOI: 10.1201/9781003285366-11

FIGURE 11.1
Participants sixth Caribbean Network Meeting, June 1999, Tobago West Indies.

In 2002, a Baltic National Group (GBa) had formed, with Krista Fischer its first National Secretary. This Group along with members in Estonia, Lativa and Lithuania merged with the Nordic Region in 2011 to form the Nordic-Baltic Region (NBR).

The British Region (BR) expanded in 2006 to include The Republic of Ireland, thus becoming the British and Irish Region (BIR). The Young Biometrician Award to a young biometrician whose education was completed within the past five years, has been awarded by BIR every two years since 2011.

Since 2006, the French Region has awarded every second year a Daniel Schwartz thesis prize, in honor of Daniel Schwartz, a key figure in founding the Société Française de Biométrie.

Elsewhere, the Caribbean Network was coalescing around the relatively new national groups Group Guatemala (GGuat) and Group Colombia (GCol), both formed in 1995. Thus, these countries along with Costa Rica, Cuba, Jamaica, and Puerto Rico, combined in 2002 to become the Central American and the Caribbean Region (RCaC), with L F Grajales as its first regional president. Figure 11.1 shows some of the participants at the June 28–30, 1999 network meeting.

On the African continent, the somewhat disparate and overlapping networks which had formed under the leadership of the British Region especially Rob Kempton as described in Section 6.8 of Chapter 6, were formally organized into a unified Sub-Saharan (SUSAN) Network in 1998 as a bilingual (French and English) Network. The first SUSAN meeting (but the sixth meeting of the earlier networks) was held in Ibadan Nigeria in August 1999. This network continued to grow; Ethiopia (GEt, founded in 1999), the Ghanian Group (GGha, 2007), the Malawi Group (GMal, 2016), and the

FIGURE 11.2
Participants at an Eastern and Southern African Network Meeting, working towards SUSAN; Rob Kempton, front row fourth from left.

Tanzanian Region (TZR, formed in 2017) were added as part of SUSAN. Many of those members who worked toward this network are shown in Figure 11.2.

In 1995, Richard Tomassone and Pierre Dagnelie had made a push for the formation of a French speaking West African network. This ultimately became the North African Region (NAR), approved in 2017, which consisted of Morocco, Algeria and Tunisia, with Hamid El Maroufy as its first president. Along the way, the Cameroon Group (GCmr) was formed in 2004; while Cameroon later dissolved as a national group, its remaining members became part of NAR as At-Large members

The Channel Network (CN) formed in 2005 includes the Belgian Region (RBe), British and Irish Region (BIR), French Region (RF) and The Netherlands Region (ANed). It meets every two years at a joint conference. After discussions that began in 2006, the Central European Network (CEN) was established in 2008, and included the Austro-Swiss Region (ROeS), German Region (DR) and Polish Region (GPol); a joint conference is held every three-four years.

Though Australasian member Norman Goodchild had first posed the idea of an Indian Subcontinent Statistics Network as early as 1986, it was not until 2018 when regions from China (GCh), India (IR), Japan (JR), Korea (RKo), Pakistan (PKSTAN, founded in 2012 following discussions begun in 2011) and Singapore (SING, founded in 2012) merged to form the East Asian Regional Network (EAR). At-Large members from Hong Kong and Taiwan are part of this network.

11.2 Publications

With regard to *Biometrics*, as reported in Chapter 7, Section 7.3, an overall Executive Editor (James Calvin) was appointed, along with the new Editor

(Raymond J Carroll, ENAR) and the new Shorter Communications Editor (Louise Ryan, ENAR), as well as an editorial assistant (Ann Hanhart). These appointments began in 1997 but naturally continued over into the following years. Still later in 1999, with the volume of submissions being too much for one editor to handle, the then (regular) Editor, Shorter Communications Editor and Executive Editor were replaced by a rotating system of three co-editors each for a three-year term with one completing his/her term in any given year and a new co-editor rotating in each year. When this new rotation was fully in place, the senior editor would be the "chair" or Coordinating Editor. With this change, the Shorter Communications section was no longer a separate section of the journal. This arrangement changed again in 2006 with the return to the appointment of an Executive Editor, and with the discontinuation of the Chair Coordinating Editor to be replaced by a Liaison Editor. In 2008, the Consultants Column was renamed Biometric Practice, with "regular" papers appearing in a renamed Biometric Methodology Section.

In 2001, the Society approved that the journal *Biometrics* became part of the JSTOR suite, found at <www.jstor.org>, soon followed by the inclusion of *Journal of Agricultural Biological and Environmental Statistics (JABES)* on this suite. This is a not-for-profit organization begun in 1995 which archives academic journals issued by professional organizations (in electronic digital form) up to five years previous to the current year. In 2003, Editorial communications such as decisions and reports began using email. The journal itself became available in electronic form to members in 2003. Discussion papers had been added in 2002; and starting in 2006, *Biometrics* authors could provide Supplementary Materials to be available on the *Biometrics* website. From 2010, all publications – *Biometrics, Biometric Bulletin, JABES* – became online publications with electronic access available to all members.

Within the regions, the Austro-Swiss Region (ROeS) teamed up with the German Region (DR) in 2004 to publish the journal *Biometrical Journal* (formerly, *Biometrische Zeitschrift*) jointly. Later, in 2017, the Chinese Region started the journal *Biostatistics and Epidemiology* with Xiao-Hua (Andrew) Zhou as Editor.

The paper Directory was replaced by a web-enabled membership Directory. Members received personalized login and passwords to access the Directory in December 2003.

11.3 Meetings

In this section, we give a brief coverage of IBCs beyond 1997. In some instances, more details are available, and so are included herein.

The first such IBC was the Nineteenth held in Cape Town in 1998. The arrangements and approvals, however, were occurring through the early-mid 1990s. Importantly, on the African continent, interest in biometry and

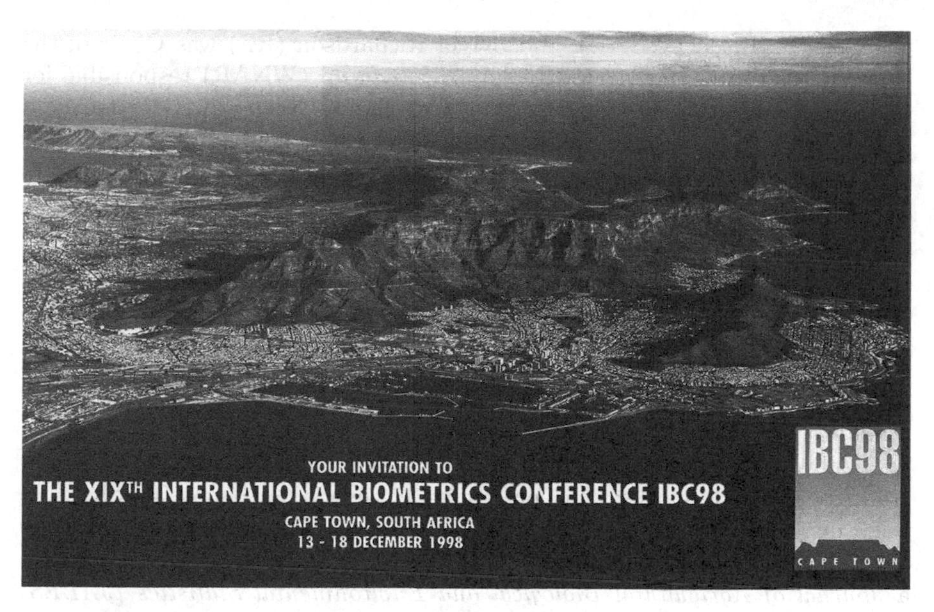

FIGURE 11.3
Flyer invitation to 1998 IBC 19 in Cape Town.

the Society's activities had been growing through the 1980–1990s primarily fostered by the British Region and spearheaded by Rob Kempton and Janet Riley (see Chapter 6, Section 6.8). At the time, members formed loosely organized networks and national groups. Thus, then-President Lynne Billard and Secretary Elisabeth Baráth (recall the Secretary was Chair of the Long Range Planning Committee which committee was responsible for making recommendations about future meeting sites; see Chapter 3) responded to invitations to attend meetings in Kampala Uganda, Ibadan Nigeria, and South Africa (at the Kruger National Park Convention Center) as well as visits with members in Kenya, Zimbabwe and Cape Town, as they grappled with these myriad desires to host an IBC. Subsequently, bids were submitted to hold the conference in Nairobi Kenya from National Group Kenya (GKe), in Harare, Zimbabwe from National Group Zimbabwe (GZim), and in Cape Town South Africa from National Group South Africa (GSAf), though importantly each National Group committed to supporting each other's bids. Ultimately, the Long Range Planning Committee recommended to Council in 1996 that the Nineteenth IBC be held in Cape Town from December 13–18, 1998. Tim Dunne was the Chair of the Local Organizing Committee; and Geoff McLachlan (AR) was Chair of the Scientific Program Committee. This was approved and announcements made in November 1996; see Figure 11.3. There were 461 registered participants.

The Twentieth IBC was held at the University of California Berkeley July 1–7, 2000, at the invitation of WNAR. The Local Organizing Committee Chair was Joan F Hilton (WNAR), and Sylvia Richardson (RF) was Chair of the Invited Program Committee with Laura Lazzeroni (WNAR) responsible for the Contributed Papers Program. President Nanny Wermuth (DR) delivered the Presidential Address "On statistical approaches common to life sciences and the social sciences." The Invited Sessions featured forty invited speakers and contributed papers constituted seventy-six parallel sessions covering 302 presentations; in addition, there were 44 poster presentations. The days of those arguments at the Cambridge IBC in 1963 about parallel sessions were now a distant memory. Attendees numbered 730.

The Twenty-first IBC was held in Freiburg Germany on July 21–26, 2002. The Scientific Program Committee was chaired by Robert N Curnow (BR), the Local Organizing Committee was chaired by Martin Schumacher (DR). The President's Address, "Are statistical contributions to medicine undervalued?," was delivered by Norman E Breslow (WNAR); like Billard's Presidential Address a decade earlier, this presentation included a poem (this time, a poem by malaria pioneer Sir Ronald Ross); see Breslow (2003). There were fifteen invited sessions, four sessions organized by the host German Region, a *Journal of Agricultural Biological and Environmental Statistics (JABES)* invited session, and fifty-seven contributed sessions (of which eight were poster sessions); there were also four one-day short courses taught before the IBC itself. The Conference venue was Albert-Ludwigs-University Freiburg; there were 649 registered participants.

For the third time in the Australasian Region, the Twenty-second IBC was held from July 11–16, 2004, in Cairns Australia, notable for its proximity to the Great Barrier Reef one of the seven great Wonders of the World, and for the fact that the *Proceedings* were distributed as a floppy disk/compact disk (CD). The Local Organizing Chair was Kaye E Basford (later 2010–2011 President), and the Scientific Program Committee Chair was Louise M Ryan (ENAR, later 2018–2019 President). The 551 attendees heard Geert Molenberghs (RBe) deliver his Presidential Address "Biometry, Biometrics, Biostatistics, Bioinformatics, . . . , Bio-X" which title alone hints at the range of topics subsumed by "biometry" since the Society's formation in 1947 (Molenberghs, 2005). The CD featured again for the Twenty-third IBC at McGill University in Montreal Canada on July 16–21, 2006. The Local Organizing Chair was James Hanley, and Geert Verbeke (RBe, later 2020–2021 President) was Scientific Program Committee Chair. There were 766 registered attendees. University College Dublin Ireland was the site for the Twenty-fourth IBC hosted by the British and Irish Region on July 13–18, 2008. The Local Arrangements Chair was John Hinde (later President 2014–2015) and the Program Committee Chair was Jean-Louis Foully (RF). Nine hundred and thirty-three people attended.

It was back to Brazil, the Region's second, for the Twenty-fifth IBC held at the Federal University of Santa Catarina in Florianópolis on December 5–10, 2010. Dalton F Andrade was the Local Arrangements Chair and Vicente Núñez-Antón (REsp) was the Program Chair. Both the Brazilan and

Argentinean Regions hosted. Kaye Basford's Presidential Address provided the membership with details of the new governance being developed for the Society; see Basford (2011). There were 820 attendees. The Twenty-sixth IBC was also held as a return visit, this time to Japan at the Kobe International Conference Center in Kobe over August 26–31, 2012. The Local Organizing Chair was Toshiro Tango, and Christine McLaren (WNAR) was the Program Committee Chair. As a first, Society membership for 2013 was included in the registration of this IBC. In another innovation, a Young Statistician Showcase was featured wherein funds to assist travel to Kobe were available for five young statisticians. Attendees numbered 622. Likewise, the Italian Region hosted for a second time when the Society was invited to hold its Twenty-seventh IBC in the Firenze Fiera Congress and Exhibition Center in Florence Italy on July 5–11, 2014. Adriano Decarli was the Local Organizing Chair and Brian Cullis (AR) was the Program Committee Chair. There were 902 registered participants.

At the invitation of WNAR, the Twenty-eighth IBC was held at the Conference Center in Victoria Canada on July 10–15, 2016. Seven hundred and nineteen attended. Laura Cowen and Youyi Fong were Local Committee Co-Chairs, while Fred van Eeuwik (ANed) was the Program Committee Chair. The Twenty-ninth IBC was hosted by the Spanish Region, its first occasion to host, at the Centre de Congressos Internacional in Barcelona Spain on July 8–13, 2018. The Local Organizing Co-Chairs were Lupe Gómez and Pere Puig; the Program Committee Chair was Charmaine Dean (WNAR). The number of attendees was 896.

The year 2020 saw the impact of the COVID virus pandemic on activities worldwide including the Thirtieth IBC. This was scheduled for Seoul South Korea from July 5 to 10, 2020; but instead occurred as a virtual conference on zoom with sessions spread across from July 6 to August 21, 2020. Despite the inherent difficulties, this IBC attracted 737 participants. While the Local Organizing Committee (Tae Rim Lee Chair, and Taesung Park Co-Chair) and Program Committee (Chair Renato Assunção, RBras) had executed their charges prior to the lock-down, the burden of mounting the virtual conference fell to the International Business Office (primarily Kristina Wolford) and the Organizing President Louise Ryan. In yet another innovation, zoom recordings of presentations were available to participants.

11.4 Governance

A major initiative since 1997 involved the massive restructuring of the Society's governance. When the decision was approved that after 2012 a region only had to have ten members, in effect eliminating the National Group terminology, the number of regions blossomed to thirty-four at the beginning of 2013. This would have a real impact on the structure of Council. Clearly, now retaining Council in its then-current format (see Chapter 10, Section 10.1) would be too unwieldy, and also finding two members from very small regions could be quite difficult to implement on a continuing basis.

Clearly, the time for a large-scale reworking of the Constitution and By-laws had arrived. Thus, in anticipation of this change, starting in February 2009 President Andrew Mead, along with immediate past president Thomas A Louis and the Executive Committee worked over the next eighteen months (in extensive consultation with Council, regions and national groups) to re-structure the Society governance culminating in an initial approval by Council in November 2010. Kaye Basford's (2011) Presidential Address at the December 2010 IBC in Florianópolis Brazil presented the new structure (without the later slight modifications necessary to conform with the 1998 Constitution; these were explained in Basford, 2013). The major change was that Council was replaced by an Executive Board (Board of Directors) consisting of three officers (President, President-elect or Outgoing President, and Secretary-Treasurer elected at large) and twelve directors (limited to two terms, four from North America, four from Europe, two from Asia/Australasia, and one each from South America and Africa) elected by members in the respective geographical areas. There would also be a Representative Council with membership elected locally, and who would serve as the conduit from the local members to the Executive Board. The number of representatives would depend on the size of the region (with one representative for regions with 10 to 49 members, two if 50 to 99, three if 100 to 499, and four if there were 500 or more members, in the respective regions[1]). Later, it was decided to ask the Representative Council Chair (elected by Council members) to attend Executive Board meetings.

These were substantial changes which would take a few years to be fully realized, so that in 2011 under Basford's leadership, the Executive Committee instituted a review of the implementation/transition process (led by Katja Ickstadt and Richardus Vonk, German Region members). The final membership approvals came in June 2012, with implementation started in January 1, 2013. Since the term "Constitution" is often synonymous with a nonprofit corporation's by-laws, having both Articles of Incorporation and a Constitution was now redundant, indeed some provisions were in conflict. Therefore, among other procedural details, the review team recommended that the Constitution (that of 1998) be merged into the By-laws. These and the Articles of Incorporation were now the Society's legal documentation. A Policies and Procedures Manual was also developed. In 2020, the "three officers" became simply "officers" allowing for the Secretary-Treasurer to be two different persons; the possibility that a treasurer might serve an additional year as an outgoing treasurer (a fifth officer) was also added if overlap with a new treasurer was needed.

---

[1]A similar structure was sent to Council for consideration in December 1964, but never adopted.

11.5 Other Actions

Until the turn of the century, elections for international officers and Council had been conducted by the Society Secretary, with nominations coming from the regions. With a professional management office in place, the International Business Office took over management of the elections. This freed the Secretary from this task and assured members of the integrity of election processes. In another move, under President Kaye Basford, voting moved in 2011 from mail-back procedures to automated electronic procedures.

In 2004, then President Geert Molenberghs initiated the formation of an ad hoc Committee to re-invigorate the earlier Education Committee. Today, the Education Committee is thriving and is concerned with the educational mission of the Society. In particular, this Committee is responsible for the Short Course Programs that have become a regular feature of IBCs. This Committee also set up the Society Journal Club, begun in 2017, as an online forum for members to discuss papers and other education initiatives.

In a different direction, although the Society and many regions had established web-pages (back in 1996), under President Louise Ryan's direction, the Society pages were reworked and a history page added.

12

For the Record

12.1 International Officers

12.1.1 President

Except for Fisher who became President in September 1947, terms are for calendar years.

Sir Ronald Aylmer Fisher (1947–1949, BR); Arthur Linder (1950–1951, AL); Georges Darmois (1952–1953, RF); William Gemmell Cochran (1954–1955, ENAR); Edmund Alfred Cornish (1956–1957, AR); Cyril Harold Goulden (1958–1959, ENAR); Léopold Martin (1960–1961, RBe); Chester Ittner Bliss (1962–1963, ENAR); David John Finney (1964–1965, BR); Luigi Luca Cavalli-Sforza (1966–1967, RItl); Gertrude Mary Cox (1968–1969, ENAR); Berthold Schneider (1970–1971, DR); Peter Armitage (1972–1973, BR); Calyampudi Radhadkrishna (CR) Rao (1974–1975, GInd); Henri Louis Le Roy (1976–1977, ROeS); John Ashworth Nelder (1978–1979, BR); Richard Melville Cormack (1980–1981, BR); Herbert Aron David (1982–1983, ENAR); Pierre Dagnelie (1984–1985, RBe); Geoffrey Harry Freeman (1986–1987, BR); Jonas Harold Ellenberg (1988–1989, ENAR); Richard Tomassone (1990–1991, RF); Niels Keiding (1992–1993, NR); Lynne Billard (1994–1995, ENAR); Byron John Treharne Morgan (1996–1997, BR); Susan Ruth Wilson (1998–1999, AR); Nanny Wermuth (2000–2001, DR); Norman Edward Breslow (2002–2003, WNAR); [Robert Alistair (Rob) Kempton, BR, elected Vice-President for 2003, died May 11 2003; Council member Kaye Basford briefly Vice-President until new elections]; Geert Molenberghs (2004–2005, RBe); Thomas Arthur Louis (2006–2007, ENAR); Andrew Mead (2008–2009, BIR); Kaye Enid Basford (2010–2011, AR); Clarice Garcia Borges Demétrio (2012–2013, RBras); John Philip Hinde (2014–2015, BIR); Elizabeth Alison Thompson (2016–2017, WNAR); Louise Marie Ryan (2018–2019, AR); Geert Verbeke (2020–2021, RBe); and José Carlos Pinheiro (2022–2023, ENAR).

12.1.2 Secretary

Chester Ittner Bliss (1947–1955, ENAR); Michael John Romer Healy (1956–1962, BR); Henri Louis Le Roy (1963–1968, ROeS); Hanspeter Thöni (1969–1975, ROeS); James S Williams (1976–1979, WNAR); Lyle David Calvin

DOI: 10.1201/9781003285366-12

(1979–1984, WNAR); Roger Mead (1985–1992, BR); Elisabeth Baráth (1993–2000, HR); Geert Molenberghs (2001–2003, RBe); Ori Davidov (2003–2005, EMR); Ashwini Mathur (2006, IR); Linda Jean Young (2007–2013, ENAR); James Carpenter (2014–2016, BIR); Bradley (Brad) John Biggerstaff (2017–2019, WNAR); Vicente Núñez-Antón (2019–2021, REsp); and Henry Mwambi (2022–2023, GSAf).

12.1.3 Treasurer

John William Hopkins (1947–1950, ENAR); Chester Ittner Bliss (1951–1956, ENAR); Allyn Winthrop Kimball (1957–1960, ENAR); Marvin Aaron Kastenbaum (1960–1963, ENAR); Henry Tucker (1963–1970, WNAR); Robert Otto Kuehl (1966, WNAR, Assistant Treasurer); Larry A Nelson (1971–1978, ENAR); Jonas Harold Ellenberg (1979–1986, ENAR); Janet Turk Wittes (1987–1990, ENAR); Stephen L George (1991–1997, ENAR); Jeffrey Thomas Wood (1998–2006, AR); Linda Jean Young (2007–2013, ENAR); James Carpenter (2014–2016, BIR); Bradley (Brad) John Biggerstaff (2017–2019, WNAR); and Vicente Núñez-Antón (2019–2023, REsp).

12.1.4 1947 Council Members

With the exception of Gertrude Cox, J B S (Jack) Haldane and Edwin B Wilson (who were elected as the first order of business when convened), the First Council consisted of those twelve individuals who constituted the International Committee of International Cooperation; specifically, Ronald Aylmer Fisher (President, UK), John William Hopkins (Treasurer, Canada), Chester Ittner Bliss (Secretary, USA), Maurice Henry Belz (Australia), Raj Chandra Bose (India), Detlev Wulf Bronk (National Research Council, USA), Gertrude Mary Cox (USA), Carlos Dieulefait (Argentina), John Burden Sanderson (Jack) Haldane (UK), Arthur Linder (Switzerland), M G Neurdenburg (The Netherlands), Georg William Rasch (Denmark), Georges Teissier (France), John Wilder Tukey (USA), Edwin B Wilson (USA). The first three listed are ex-officio Council members by virtue of their office. The Second Council meeting of September 15 1947, elected seven more members; namely, Adriano Buzzati-Traverso (Italy), Kenneth Stewart Cole (USA), Milislav Demerec (USA), Cyril Harold Goulden (Canada), Ivor J Johnson (USA), Joseph Needham (France), and George Gaylord Simpson (USA).

12.2 Regional and National Group Officers

For the record, we provide a list of regional presidents and/or national secretaries, as appropriate. Prior to 1954, the regional presidents were international Vice-Presidents after which, following a 1953 Council decision,

TABLE 12.1
Region (R), National Group (G), and Network (N) Abbreviations

RArg	Argentinian	RItl	Italian
AR	Australasian	JR	Japanese
ROeS	Austro-Swiss	GKe	Kenyan
GBa	Baltic	RKo	Korean
RBe	Belgian	GMal	Malawi
GBot	Botswanian	GMex	Mexico
RBras	Brazilian	GNa	Namibia
BR/BIR	British/British and Irish	GNi	Nigerian
GCmr	Cameroon	ANed	The Netherlands
RCaC	Central American and Caribbean	NAR	North African
GCl	Chilian	NR/NBR	Nordic/Nordic-Baltic
GCh	China	GNo	Norway
GCol	Colombia	PKSTAN	Pakistani
GDe	Danish	GPol	Polish
EMR	Eastern Mediterranean	GRo	Romanian
ENAR	Eastern North American	SING	Singaporean
ECU	Ecuadorian	GSAf	South Africa
GEth	Ethopian	REsp	Spanish
RF	French	GSD	Sweden
DR	German/RGDR East German	TZR	Tanzanian
GGha	Ghanian	GUgan	Ugandan
GGuat	Guatemala	WNAR	Western North American
GHU/HR	Hungary	GZim	Zimbabwean
IR/GInd	Indian	AL	At-Large
NGi	Indonesia		
Networks:			
CEN	Central European	EAR	East Asian Regional
CN	Channel	SUSAN	Sub-Saharan

they were officially referred to as the "Regional President." Abbreviations for these regional and national groups are given in Table 12.1.

12.2.1 Argentina Region (RArg)

1999–2001 - Nélida Winzer; 2002–2005 - Monica G Balzarini; 2006–2009 - S J Bramardi; 2010–2015 - Julio Di Rienzo; 2016–2017 - María Gabriela Cendoya; 2020–2021 - Pablo Daniel Reeb.

12.2.2 Australasian Region (AR)

Vice-President of Society: 1948–1949, 1950 - Edmund Alfred (Alf) Cornish; 1951–1953 - Clifford (Cliff) W Emmens; Regional President of AR: 1953–

1954 - Helen Newton Turner; 1955–1957 - Evan James Williams; 1958–1959 - Clifford (Cliff) W Emmens; 1960–1962 - Maurice Henry Belz; 1963–1965 - J Henry Bennett, 1966–1967 - Evan J Williams; 1968 - James (Jim) Bartram Douglas; 1969–1970 - Peter J Claringbold; 1971–1973 - Stephen Lipton; 1974–1975 - Henry M Finucan; 1976–1977 - W Bruce Hall; 1978–1979 - Ronald L Sandland; 1980–1981 - K P Haydock; 1982–1983 - Jeffrey (Jeff) Thomas Wood; 1984–1985 - Ivor S Francis 1986–1987 - Susan Ruth Wilson; 1988–1989 - John N Darroch; 1990–1992 - Charles A McGilchrist; 1993–1996 - J A (Nye) John; 1997–1998 - Kaye Enid Basford; 1999-2000 - Ken G Russell; 2001–2002 - Peter Johnstone; 2003–2004 - John Reynolds; 2005–2006 - Ann Cowling; 2007–2008 - Melissa Dobbie; 2009–2010 - Graham Hepworth; 2011–2012 - Mario D'Antuono; 2013–2014 - David Baird; 2015–2016 - Ross Darnell; 2017–2018 - Samuel Müller; 2019–2020 - Alan Hepburn Welsh; 2021–2022 - Vanessa Cave.

12.2.3 Region Österreich-Schweiz (Austro-Swiss Region, ROeS)

1962–1963 - Leopold Karl Schmetterer; 1964–1965 - Arthur Linder; 1966–1967 - Alois Kaelin; 1968–1969 - Franz Xaver Wohlzogen; 1970–1971 - Henri Louis Le Roy; 1972–1973 - Eugen Olbrich; 1974–1975 - Hans Riedwyl; 1976–1977 - E Lengauer; 1978–1979 - F H Schwazenbach; 1980–1981 - Peter Bauer; 1982–1983 - Uwe Ferner; 1984–1985 - Victor Scheiber; 1986–1987 - Hugo Flühler; 1988–1989 - Josef Gölles; 1990–1992 - Christoph E Minder; 1993–1994 - A Neiss; 1994–1995 - Hans-Rudolf Roth; 1996–1997 - Michael Schemper; 1998–1999 - Eric Lüdin; 2000–2001 - Karl P Pfeiffer; 2002–2003 - Willi Maurer; 2004–2005 - Andrea Berghold; 2006–2007 - Hans Ulrich Burger; 2008–2009 - Martina Mittlböck; 2010–2011 - N Neumann; 2012–2013 - H Ulmer; 2014–2015 - Leonhard Held; 2016–2017 - Martin Posch; 2018–2019 - Valentin Rousson; 2020–2021 - Arne Bathke.

12.2.4 Région Belge (Belgian Region, RBe)

Vice-President of Society: 1953 - Paul Spehl; Regional President of RBe: 1954 - M Simon; 1955 - E Cordiez; 1956 - Desiré DeMeulemeester; 1957–1958 - R Laurent (died in office April 1958); 1958 - P P Denayer; 1959–1960 - Jean-M A G Henry; 1961–1962 - Maurice Welsch; 1963–1964 - Jean-M A G Henry; 1965 - Paul Spehl (Honorary President); 1965–1966 - A Vanden Hende; 1967–1968 - R Consael; 1969–1970 - Léopold Martin; 1971–1972 - Roger Bontemps; 1973–1974 - Pierre Dagnelie; 1975–1976 - Paul Berthet; 1977–1978 - A H L Rotti; 1979–1980 - Guy Gérard; 1981–1982 - Jean-Jacques Claustriaux; 1983–1984 - Robert Oger; 1985–1986 - Paul Berthet; 1987–1988 - Ernest Feytmans; 1989–1991 - Jean-Jacques Claustriaux; 1991–1992 - Éric Le Boulengé; 1993–1994 - Robert Oger; 1995–1998 - Herman H Calleart; 1997–1998 - Adelin Albert; 1999-2001 - Els Goetghebeur; 2001–2002 - Marc Buyse,

2002–2003 - Geert Verbeke; 2004–2006 - W Malbecq; 2006–2007 - L Bijnens; 2008–2009 - A Robert; 2010–2012 - L Duchateau; 2013–2015 - C Legrand; 2014–2017 - An Vanderbosch; 2016–2017 - Sophie Vanbelle; 2017–2018 - Sophie Vanbelle; 2018–2019 - Olivier Thas; 2019–2021 - Pierre Lebrun.

12.2.5 Região Brasileira (Brazilian Region, RBras)

1956–1958 - Constantino Gonçalves Fraga Jr; 1958–1960 - Frederico Pimental Gomes; 1960–1962 - Adolpho Martins Penha; 1963–1964 - Paulo Mello Freire; 1965 - Constantino Gonçalves Fraga Jr; 1966–1968 - Frederico Pimental Gomes; 1968–1969 - Elza S Berquó; 1970 - Adelpho Martins Penha; 1971–1973 - Armanda Conagin; 1974–1975 - Frederico Pimental Gomes; 1976–1977 - Roland Vencovsky; 1978–1979 - Aldir Alves Teixeira; 1980–1981 - Frederico Pimental Gomes; 1982–1983 - João Gilberto Corrêa da Silva; 1984–1985 - Aldir Alves Teixeira; 1986–1987 - Décio Barbin; 1988–1989 - Clóvis de Araújo Peres; 1990–1992 - Clarice Garcia Borges Demétrio; 1993–1994 - João Gilberto Correa da Silva; 1994–1996 - Sérgio do Nascimenta Kronka; 1996–1998 - Carlos Roberto Padovani; 1998–2002 - Clarice Garcia Borges Demétrio; 2000–2002 - Maria Cecilia Mendes Barreto; 2002–2004 - Clarice Garcia Borges Demétrio; 2004–2005 - Joel Augusto Muniz; 2006–2010 - Paulo Justiniano Ribeiro; 2011–2013 - Roseli A Leandro; 2014–2016 - Paulo Justiniano Ribero Jr.; 2017–2018 - Alessandro Dal'Col Lúcio; 2019 - Joel Augusto Muniz; 2020–2021 - Paulo Jorge Canas Rodrigues.

12.2.6 British Region (BR)/British and Irish (BIR, 2006)

Vice-President of Society: 1948–1949 - John William Trevan; 1950–1951 - Ronald Aylmer Fisher; 1952–1953 - Frank Yates; Regional President of BR: 1954–1955 - Robert Russell Race; 1956–1957 - David John Finney; 1958–1959 - Joseph Oscar Irwin; 1960 - Kenneth Mather; 1961–1962 - John Alexander Fraser Roberts; 1963–1964 - John Henry Gaddum (Sir John from 1964); 1965–1966 - Maurice Stevenson Bartlett; 1967–1968 - John Gordon Skellam; 1969–1970 - Stanley Clifford (Clifford) Pearce; 1971–1972 - Cedric A B Smith; 1973–1974 - P S Hewlett; 1975–1976 - Peter D Oldham; 1977–1978 - C David Kemp; 1979–1980 - Robert N Curnow; 1981–1982 - Geoffrey Harry (Geoff) Freeman; 1983–1984 - Philip Holgate; 1985–1986 - John N R Jeffers; 1987–1988 - Peter Armitage; 1989–1990 - John Clifford Gower; 1990–1992 - Richard Melville Cormack; 1993–1994 - Anthony William Fairbanks Edwards; 1995–1996 - Robert (Rob) Alistair Kempton; 1997–1998 - Doug Altman; 1999–2000 - Tom B L Kirkwood; 2001–2002 - Rosemary Anne Bailey; 2002–2004 - Michael G Kenward; 2005–2006 - Joe N Perry; 2007–2008 - David Balding; 2009–2010 - Byron John Treharne Morgan; 2011–2012 - John Philip Hinde; 2013–2014 - Simon Thompson; 2015–2016 - John Matthews; 2017–2018 - Martin Ridout; 2019–2020 - Ruth King; 2021 - Daniel Farewell.

12.2.7 Central American and Caribbean (RCaC)

2002–2003 - L F Grajales; 2004 - J C Sandoval; 2005 - J Camacho; 2006–2007 - Raúl Macchiavelli; 2008–2009 - Bruce Lauckner; 2010–2013 - J R Demey; 2014–2015 - L Lopez; 2016–2017 - L Perez; 2020–2021 - Luis Alberto López.

12.2.8 Eastern Mediterranean Region (EMR)

2001–2003 - Laurence S Freedman; 2003–2005 - Hüseyin Refik Burgut; 2006–2007 - Urania Dafni; 2007–2009 - David M Zucker; 2010–2011 - Ergun Karaagaoglu; 2011–2013 - Stergios Tzortzios; 2013–2015 - Havi Murad; 2015–2016 - Ilker Unal; 2017–2018 - Christos Nakas; 2019–2020 - Itai Dattner; 2021 - Yoav Benjamini; 2022 - Konstantinos Fokianos.

12.2.9 Eastern North American Region (ENAR)

Vice-President of Society: 1948–1949 - Charles Paine Winsor; 1950 - Joseph Berkson; 1951–1952 - Horace W Norton; 1953 - S Lee Crump; Regional President of ENAR: 1954 - S Lee Crump; 1955–1956 - David B Duncan; 1957–1958 - Boyd Harshbarger; 1959 - Jerome J Cornfield; 1960 - Walter T Federer; 1961 - Oscar Kempthorne; 1962 - Henry L (Curly) Lucas; 1963 - Herbert Aron David; 1964 - Theodore Alfonso (Ted) Bancroft; 1965 - Ralph Allan Bradley; 1966 - Richard Loree Anderson; 1967 - Paul Meier; 1968 - Herman Otto (HO) Hartley; 1969 - Samuel W Greenhouse; 1970 - Douglas S Robson; 1971 - Bernard G Greenberg; 1972 - Edmund A Gehan; 1973 - James E Grizzle; 1974 - Charles W Dunnett; 1975 - Richard G Cornell; 1976 - Foster B Cady; 1977 - Nathan Mantel; 1978 - Yvonne Millicent Mahala Bishop; 1979 - Gary Koch; 1980 - Charles E Gates; 1981 - Jesse C Arnold; 1982 - Theodore Colton; 1983 - S Michael Free Jr; 1984 - Peter A (Tony) Lachenbruch; 1985 - Lynne Billard; 1986 - Joseph L Fleiss; 1987 - John R Van Ryzin (Died in office, March 1987)/Joseph L Fleiss; 1988 - Mitchell Gail; 1989 - James H Ware; 1990 - Stephen Lagakos; 1991 - Barbara Tilley; 1992 - Thomas Arthur Louis; 1993 - David A DeMets; 1994 - Judith Rich O'Fallon; 1995 - Janet Turk Wittes; 1996 - Scott L Zeger; 1997 - Judith D Goldberg; 1998 - Joel Greenhouse; 1999 - Susan Smith Ellenberg; 2000 - Louise Marie Ryan; 2001 - Linda Jean Young; 2002 - Carol K Redmond; 2003 - Timothy G Gregoire; 2004 - Marie Davidian; 2005 - Peter Emrey; 2006 - Jane Pendergast; 2007 - Lisa LaVange; 2008 - Eric (Rocky) Feuer; 2009 - Lance Waller; 2010 - Sharon Lise Normand; 2011 - Amy Herring; 2012 - Karen Bandeen-Roche; 2013 - Daniel Heitjan; 2014 - DuBois Bowman; 2015 - José C Pinheiro; 2016 - Jianwen Cai; 2017 - Scarlett Bellamy; 2018 - Jeffrey S Morris; 2019 - Sarah Ratcliffe; 2020 - Mike Daniels; 2021 - Brent Coull; 2022 - Simone Gray.

12.2.10 Région Française (French Region, RF)

Vice-President of Society: 1949–1951 - Maurice René Fréchet; 1952–1953 - Georges Teissier; Regional President of RF: 1954–1955 - Georges Darmois; 1956–1957 - Eugene Morice; 1958–1959 - André Vessereau; 1960–1961 - Ph L'Heritier; 1962–1963 - Daniel Dugué; 1964–1965 - Daniel Bargeton; 1966 - Sully Ledermann; 1967–1970 - Jean Sutter; 1971–1973 - Daniel Schwartz; 1974–1982 - Jean Marie Legay; 1983–1985 - Joseph Lellouch; 1986–1989 - Jean Pierre Masson; 1990–1992 - Bernard Asselain; 1993–1994 - Jean Tranchefort; 1995–1996 - Richard Tomassone; 1997–1998 - Michel Chavance; 1999-2000 - Bruno Falissard; 2000–2002 - Avner Bar-Hen; 2003–2004 - Bernard Asselín; 2005–2008 - Robert Faivre; 2009–2010 - Bernard Asselaín; 2011–2012 - Isabelle Albert; 2013–2014 - Pascal Pierre Wild; 2015–2017 - Daniel Commenges; 2018–2020 - Mounia N Hocine; 2021 - David Causeur.

12.2.11 Deutsche Region (German Region DR)

1955–1956 - Egon Ullrich; 1957–1958 - Ottokar Heinisch; 1959–1960 - A Augsberger; 1961 - Ottokar Heinisch; 1962–1963 - J Hartung; 1964–1965 - Siegfried Koller 1966 - vacant (Herbert Jordan); 1967–1968 - Berthold Schneider; 1969–1970 - Hans Rundfeldt; 1971 - E Walter; 1972–1973 - Rainald K Bauer; 1974–1975 - Hans Klinger; 1976–1977 - Gustav A Lienert; 1978 - H Fink; 1979 - R Repges; 1980–1981 - Hanspeter Thöni; 1982–1983 - R Repges; 1984–1985 - E Sonnemann; 1986–1988 - Berthold Schneider; 1989–1990 - Siegfried Schach; 1991 - H J Bernd Streitberg (died in office September 1991); 1992 - Rolf J Lorenz; 1993 - Max P Baur; 1994 - Hanspeter Thöni; 1995–1996 - Nanny Wermuth; 1997 1998 - Joachim Vollmar; 1999–2000 - Iris Pigeot-Kuebler; 2001–2003 - Guido Giani; 2003–2004 - Joachim Röhmel; 2005–2006 - Andreas Ziegler; 2007–2009 - Ludwig Hothorn; 2010–2011 - Richardus Vonk; 2012 - Katja Ickstadt; 2013–2014 - Jürgen Kübler; 2014–2016 - Tim Friede; 2017–2019 - Andreas Faldum; 2020 - Werner Brannath; 2021 - Annette-Kopp Schneider.
Deutsche Democratic Republic Region (German Democratic Republic Region (RGDR), East German Region (GDR): 1971–1972 - Joachim-Hermann Scharf; 1973–1974 - Johannes Adam; 1975–1976 - Habil Dieter Rasch; 1977–1979 - Gottfried Enderlein; 1980–1981 - Heinz Ahrens; 1982–1983 - Gottfried Enderlein; 1984–1986 - Heinz Ahrens; 1987–1989 - Helmut Enke; 1990–1991 - Klaus-D Wernecke.

12.2.12 Hungary Region (HR)

1989–1994 - Elisabeth Baráth; 1995–1997 - Zsolt Harnos; 1998–2002 - Béla Tóthmérész; (Region dissolved 2003, reverted back to Group).

12.2.13 Indian Region (IR)

Vice-President of Society: 1949–1950 - Prasanta Chandra Mahalanobis; (Region dissolved 1951; re-formed 1989); Regional President of IR: 1989–1997 - Pandurang Vasudeo Sukhatme; 1997–1998 - Prem Narin; 1999–2004 - Girja Kant Shukla; 2005–2008 - H Sridhara; 2009–2012 - V G Kaliaperumal; 2013–2020 - Perumal Venkatesan; 2021 - Ajit Sahai.

12.2.14 Regione Italiana (Italian Region, RItl)

Vice-President of Society: 1953 - Claudio Barigozzi; Regional President of RItl: 1953–1955 - Claudio Barigozzi; 1956–1957 - Gustavo Barbensi; 1958–1961 - G Montalenti; 1962–1964 - Gustavo Barbensi; 1965-1968 - Luigi Luca Cavalli-Sforza; 1969–1972 - Adriano Buzzati-Traverso; 1973–1974 - Gustavo Barbensi (died in office 1974); 1975–1977 - Giulio Alfredo Maccacaro (died in office January 1977); 1978–1980 - Giorgio Segre; 1981–1983 - Claudio Barigozzi; 1985–1989 - Italo Scardovi; 1990–1991 - Ercole Ottaviano; 1992–1995 - Ettore Marubini; 1996–1999 - Alberto Piazza; 2000–2003 - Carla Rossi; 2004–2007 - Annibale Biggeri; 2007–2013 - A De Carli; 2014–2016 - M G Valsecchi; 2017–2019 - Clelia Di Serio; 2020–2021 - Stefania Galimberti.

12.2.15 Japanese Region (JR)

1979–1986 - Chikio Hayashi; 1987–1988 - Tadakazu Okuno; 1989–1993 - Akira Sakuma; 1994–1995 - Tsutomu Komazawa; 1996–2000 - Isao Yoshimura; 2001–2004 - T Yanagawa; 2005–2008 - Toshiro Tango; 2009–2012 - Tosiya Shun Sato; 2013 - Yasuo Ohashi; 2014 - Tosiya Shun (Shun) Sato; 2015–2017 - Yasuo Ohashi; 2020–2021 - Shigeyuki Matsui.

12.2.16 Korean Region (RKo)

2001 - Byung Soo Kim; 2001–2003 - Hyonggi Jung; 2004–2012 - Tae Rim Lee; 2013–2014 - Taesung Park; 2015–2016 - D Kim; 2017 - S Lee; 2020 - Sangbum Choi; 2021 - Jaehee Kim; 2022 - Sohee Park.

12.2.17 Afdeling Netherland (The Netherlands Region, ANed)

1960–1962 - David Karel de Jongh (died in office 1962); 1963–1975 and 1979-82 - (no one – Secretary-Treasurer (1960-82) Hendrik de Jonge stepped in); 1976–1978 - Leo Casper Anton Corsten; 1986–1993 - Roel van Strik; 1994–1996 - Johannes (Hans) Cornelis van Houwelingen; 1997–2000 - Theo Stijnen; 2001–2005 - Alfred Stein; 2006–2008 - Aeilko H Zwinderman; 2009–2013 - Fred van Eeuwijk; 2014–2017 - Jeanine Houwing-Duistermaat; 2017–2018 - Ernst C Wit; 2018–2021 - Mark van de Wiel.

12.2.18 Nordic Region (NR)/Nordic Baltic (NBR, from 2011)

1982–1983 - Aage Vølund; 1984–1985 - Hans Wedel; 1986–1987 - Timo Hakulinen; 1988–1989 - Holmgeir Björnsson; 1990–1991 - Jon Stene; 1992–1993 - Juni Palmgren; 1994–1996 - Odd O Aalen; 1997–2001 - Sture Holm; 2001–2005 - Mikael Vaeth; 2006–2008 - Esa Läärä; 2009–2012 - Geir Egil Eide; 2013–2016 - Krita Fischer; 2017–2020 - Ziad Taib; 2021 - Andreas Kryger Jensen.

12.2.19 Spanish Region (REsp)

1993–1995 - Carmen Santisteban Requena; 1996 - Carles M Cuadras; 1997–1998 - Emilio Carbonell; 1999–2000 - Guadalupe Gómezi Melis; 2001–2003 - María-Jesús Bayarri; 2004–2005 - José Luis González Andújar; 2006–2007 - Juan Luis Chorro Gascó; 2008–2009 - Álex Sánchez Pla; 2010–2011 - Carmelo Ávila-Zarza; 2012–2013 - Maria Luz Calle Rosignana; 2014–2015 - David V Conesa Guillén; 2016–2017 - Immaculata Arostegui; 2018–2019 - Klaus Langohr; 2020–2021 - Pere Puig.

12.2.20 Western North American Region (WNAR)

Vice-President of Society: 1948–1950 - Frank Walter Weymouth; 1951–1952 - G A Baker; 1953 - Blair M Bennett; Regional President of WNAR: 1954 - Douglas G Chapman; 1955 - Wilfred J Dixon; 1956–1957 - Douglas G Chapman; 1958 - Joseph L Hodges Jr.; 1959–1961 - William F Taylor; 1962–1963 - Carl A Bennett; 1964–1965 - Lyle David Calvin; 1966–1967 - Stanley W Nash; 1968 - Carl E Hopkins; 1969 - Lincoln Moses; 1970 - Olive Jean Dunn; 1971 - Donovan J Thompson; 1972 - Edward B Perrin; 1973 - James S Williams; 1974 - Virginia A Clark; 1975 - Robert Otto Kuehl; 1976 - Robert M Elashoff; 1977 - Byron W (Bill) Brown; 1978 - N Scott Urquhart; 1979 - Calvin Zippin; 1980 - Donald Guthrie; 1981 - Thomas J Broadman; 1983 - Gerald van Belle; 1983 - Roger G Peterson; 1984 - Ross L Prentice; 1985 - Lyman L McDonald; 1986 - Jessica Utts; 1987 - John J Crowley; 1988 - J Richard Alldredge; 1989 - Loveday L Conquest; 1990 - R Kirk Steinhorst; 1991 - Nick P Jewell; 1992 - Barbara McKnight; 1993 - Terrance P Speed; 1994 - Alice Whittemore; 1995 - Stephanie Green; 1996 - David L Turner; 1997 - Joan F Hilton; 1998 - Elizabeth Alison Thompson; 1999 - Scott Emerson; 2000 - Denise J Roe; 2001 - Edward John Bedrick; 2002 - Charmaine Dean; 2003 - Anna Barón; 2004–2005 - Wesley Johnson; 2006 - Christine McLaren; 2007 - Kenneth Paul Burnham; 2008 - John Neuhaus; 2009 - Todd Alonzo; 2010 - Peter A (Tony) Lachenbruch; 2011 - Dan Gillen; 2012 - John Kettelson; 2013 - Bradley (Brad) John Biggerstaff; 2014 - Elizabeth Brown; 2015 - Catherine M Crespi; 2016 - Susanne May; 2017 - Sarah Emerson; 2018 -

Motomi Mori; 2019 - Ying Lu; 2020 - Katerina Kechris; 2021 - Ying Lu; 2022 - GaryChan.

12.2.21 National Secretaries

From 2013, national groups became regions regardless of size, with the National Secretary as its chief officer, de facto "president."

Argentina (GArg): 1996–1997 - R B Ronceros; 1997–1998 - Nélida Winzer; (Became Region 1999).

Baltic (GBa): 2003–2010 - Krista Fischer; (Merged with Nordic Region 2011).

Belgian (GBe): 1952–1953 - Léopold Martin; (Became Region 1953).

Botswanna (GBot): 1994 - Manibhai S Patel; 1995–1997 - D K Shah; 1998–2004 - S R T Moeng; 2005–2007 - P M Kgosi; 2008–2021 - Njoku Ola Ama.

Brazil (GBras): 1954–1956 - Américo Groszmann; (Became Region 1956).

Cameroon (GCmr): 2004–2014 - Innocent Zebaze.

Chile (GCl): 2001–2002 - M G Icaza Noguera; 2003–2004 - G Marshall; 2005–2010 - C Silva; 2011–2016 - C Meza; 2017 - L M Castro; 2020–2021 - Jorge Figueroa-Zúñiga.

China (GCh): 1987–2009 - Ji-Qian Fang; (Reactivated 2012); 2012–2021 - Xiao-Hua (Andrew) Zhou.

Colombia (GCol): 1995–1996 - P Pacheco; 1997–2001 - A Perez; (Became part of Central American-Caribbean Region 2002).

Deutsche Democratic Republic, (East German, GGDR): 1969–70 - Erna Weber; 1971 - Johannes Adam; (Became Region 1971).

Denmark (GDe): 1951–80 - Niels F Gjeddebaek; 1981 - Aage Vølund; (Became part of Nordic Region 1982).

Ecuadorian (ECU): 2015–21 - Omar Honorio Ruíz Barzola.

Ethiopia (GEth): 1999–2015 - Girma Aweke Taye; December 2015–21 - Anteneh Tessema Yalew.

German (GDR): 1952–54 – Maria-Pia Geppert; (Became Region 1954).

Ghana (GGha): 2007–2019 Agnes Ankomah; 2019–2021 - Abukari Alhassan; 2022 - N Forcheh.

Guatemala (GGuat): 1995–1999 - Jorge Matute; 2000–2001 - V Alvarez; (Became part of Central American and Caribbean Region 2002).

Hungary (GHU): 1965–1980 - Ireneusz Juvancz; 1981–1986 - János Sváb (died in office May 1986); 1986–1988 - Elisabeth Baráth; (Region 1989–2003); 2004 - Zsolt Harnos; 2005–10 - L Hufnagel; (Inactive since 2011).

India (GInd): 1952–1955 - Vinayak Govind Panse; 1956–1959 - K Kishen; 1960–1963 - A R Roy; 1964–1967 - A R Kamat; 1968–1969 - G R Seth; 1970–1972 - M S Chakrabarti (died suddenly in office 1972); 1972–1977 - Y S Sathe; 1977–1989 - Girja Kant Shukla (Returned to Region 1989).

Indonesia (NGi): 1972–1991 - Andi Hakim Nasoetion; 1993–2009 - H Aunuddin; (Inactive since 2010).

Italy (GItl): 1949–1950 - Adriano Buzzati-Travserso; 1951–1952 - Luigi Luca Cavalli-Sforza; (Became Region 1953).

Japan (GJap): 1953–1970 - Matayoshi Hatamura; 1971–1978 - Kenziro Saio; (Became Region 1979).

Kenya (GKe): 1989–1991 - Manibhai S Patel; 1992–1993 - Sagary Nokoe; 1993–1995 - Hezron Oranga; 1996–2012 - John M Odhiambo; 2013–2014 - R Nguti; 2015–2020 - John Ali Mwangi; 2021 - Henry Kissinger Ocgieng Athiany.

Korea (GKo): 1989 - Sung H Park; 1990–1992 - Han-Poong Shin; 1993–1995 - W Hahn; 1996–1997 - Seung Wook Lee; 1998–2000 - Byung Soo Kim; (Became Region 2000).

Malawi (GMal): 2016–2021 - Jupiter Simbeye.

Mexico (GMex): 1967–78 - Eduardo Casas Díaz; 1979–1980 - Alberto Castillo Morales; 1981–2002 - Angel Martínez Garza; 2003–2009 - H Vaquera Huerta; (Became part of Central American and Caribbean Region 2009).

Morocco: (GMo) 1989–2010 - Ahmed Goumari; (Became part of North African Region 2017).

Namibia (GNam): 2001–2010 - N Ola Ama.

The Netherlands (GNed): 1949–1953 - M G Neurdenburg; 1953–59 - E van der Laan; 1960–1961 - Hendrik de Jonge; (Became Region 1962).

Nigeria (GNi): 1997–2000 - Sagary Nokoe; 2001–2014 - Ben A Oyejola; 2015–2017 - Osebekwin E Asiribo; 2020–2021 - Emmanuel Tejujola Jolayemi.

North Africa (NAR): 2017–2021 - Hamid El Maroufy.

Norway (GNo): 1960–1981 - Lars K Strand; (Became part of Nordic Region 1982).

Pakistani (PKSTAN): 2012 - Muhammad Aslam; 2021 - Alia Sajjad.

Poland (GPol): 1992–2001 - Anna Bartkowiak; 2001–2018 - Stanislaw Mejza; 2019–2021 - Małgorzata Graczyk; 2022 - Stanislaw Mejza.

Romania (GRo): 1970–2000 - Tiberiu Postelnicu; 2001–2017 - Cornelia Enãchescu.

Singapore (SING): 2012–2021 - Jialiang Li.

South African Group: (GSAf) 1991 - Arthur Asquith Rayner; 1992–1993 - Tim Dunne; 1994–1996 - John H Randall; 1997–2000 - Harvey Morgan Dicks; 2001–2006 - Peter M Njuho; 2007–2013 - Freedom N Gumedze; 2014–2021 - Iain MacDonald.

Spanish (GSp): 1985–1991 - Carmen Santisteban Requena; (Became Region 1992).

Sweden (GSD): 1954–1965 - Herman O A Wold; 1966–1970 - Bertil Matéern; 1970–1972 - Adam Taube; 1973–1976 - K G Jöreskog; 1977–1978 - Hans Wedel; 1979–1981 - Paul Seeger; (Became part of Nordic Region 1982).

Switzerland: 1954–1956 - Arthur Linder, 1956–1961 - Henri L Le Roy; (Became ROeS 1962).

Tamzanian (TZR): 2017–2021 - Innocent B Mboya.

Uganda (GUgan): 1994–2000 - Leonard K Atuhaive; 2001–2010 - M Nabasirye; 2012–2021 - Dan K Kajunga.

Urugary (GUru): 1999–2009 - Ramon Alvarez Vaz; 2012–2020 - Dan K Kajungu.

Venezuela (GVen): 1995–2001 - Laura E Pla; (Became part of Central American and Caribbean Region 2002).

Zimbabwean (GZim): 1991–1994 - Jane Canhão; 1995–1998 - Erica C Keogh; 1999–2014 - Bernard Chasekwa; 2015–2020 - Betty Mawire; 2021 - Cherlynn Simbiso Dumbura.

12.3 Editorial Officers

We provide a list of editors for the respective journals. Editors for the subsections, Queries and Notes, and Book Reviews for *Biometrics* are also included. Recall the regional abbreviations can be found in Table 12.1.

12.3.1 Editors of *Biometrics*

Editors to 1998 had fixed terms as the single editor, after which sets of three editors are co-editors on staggered three-year terms.

Gertrude M Cox (1945–55, ENAR); John W Hopkins (1956–57, ENAR); Ralph A Bradley (1957–62, ENAR); Michael R Sampford (1962–67, BR); Herbert A David (1967–72, ENAR); Frank A Graybill (1972–75, WNAR); Foster B Cady (1975–79, ENAR); Peter Armitage (1979–84, BR); Daniel L Solomon (1985–89, ENAR); Klaus Hinkelmann (1989–93, ENAR); Charles A McGilchrist (1994–97, AR); Raymond J Carroll (February 1997–April 2000, ENAR); Louise M Ryan (1999–January 2000, ENAR); Anthony N Pettitt (November 1999–December 2001, AR); Marie Davidian (February 2000–December 2002, ENAR); Daniel Commenges (May 2000–03, RF); Brian Cullis (2002–04, AR); Xihong Lin (2003–05, ENAR); Mike Kenward (2004–06, BR); Laurence S Freedman (2005–07, EMR); Naisyin Wang (2006–08, ENAR); Geert Molenberghs (2007–09, RBe); David M Zucker (2008–10, EMR); Thomas A Louis (2009–11, ENAR); Geert Verbeke (2010–12, RBe); Russell Millar (2011–13, AR); Jeremy M G Taylor (2012–14, ENAR); Jeanine Houwing-Duistermaat (2013–15, ANed); Yi-Hau Chen (2014–16, GCh); Michael J Daniels (2015–17, ENAR); Stijn Vansteelandt (2016–18, RBe); Malka Gorfine (2017–19, EMR); Debashis Ghosh (2018–20, WNAR); Mark Brewer (2019–21, BIR); Alan H Welsh (2020–22, AR). Executive Editor: James A Calvin (1997–99, ENAR); Marie Davidian (2006–17, ENAR); Geert Molenberghs (2018–, RBe).

12.3.1.1 Queries and Notes; Shorter Communications

Prior to 1974, this section was "Queries and Notes"; it became "Shorter Communications" in 1974, until 1999. George W Snedecor (1945–58, ENAR); David J Finney (1959–62, BR); John A Nelder (1962–66, BR); Peter Sprent (1967–71, BR); C David Kemp (1972–75, BR); James S Williams (1976–79, WNAR); John J Gart (1979–84, ENAR); Robin Thompson (October 1984–88, BR); Niels Keiding (1989–92, NR); Byron J T Morgan (1993–September 1997, BR); Louise M Ryan (December 1997–99, ENAR).

12.3.1.2 Book Reviews

John G Skellam (1960–63, BR); Walter T Federer (1964–March 1972, ENAR); FN David (1972–77, WNAR); Richard M Cormack (1978–84, BR); C David Kemp and Freda Kemp (1984–March 2000, BR); Martin Ridout (June 2000–03, BR); Iris Pigeot-Kübler (2004–05, DR); Thomas M Loughin (2006–09, ENAR); Guilherme J M Rosa (2009–March 2013, RBras); Taesung Park (June 2013–15, RKo); Donna Pauler Ankerst (2016–18, DR); Brisa N Sánchez (2019–21, ENAR); Chuhsing Kate Hsiao (2021–, EAR).

12.3.2 Editors of *Biometric Bulletin*

Robert O Kuehl (1984–87, WNAR); Joe N Perry (1988–February 1992, BR); Girja Kant Shukla (May 1992–February 1995, IR); Richard Tomassone (May 1994–December 1997, RF); Geert Molenberghs (1998–2000, RBe); Ton Ten Have (2001–04, ENAR); Urania Dafni (2005–08, EMR); Roslyn A Stone (2009–12, ENAR); Dimitris Karlis (2013–15, EMR); Havi Murad (2016–18, EMR); and Ajit Sahai (2019–21, IR).

12.3.3 Editors of *Journal of Agricultural Biological and Environmental Statistics (JABES)*

Dallas E Johnson (1994–98, ENAR); Bryan F J Manly (1999–2001, AR); Linda J Young (2002–04, ENAR); Byron J T Morgan (2005–07, BR/BIR); Carl Schwarz (2008–10, WNAR); Montserrat (Montse) Fuentes (2011–15, ENAR); Stephen T Buckland (2016–18, BIR); Brian Reich (2019–21, ENAR).

12.3.4 Regional Journals

12.3.4.1 *Biométrie-Praximétrie*, RBe

Léopold Martin (1960–63); Pierre Dagnelie (1964–68); Paul Berthet (1969–70); Philippe Smets (1971–75); Jacques Delincé (1976–78); Jean-Jacques Claustriaux and Robert Oger (1979–85); Guy Bouxin and Robert Oger (1986–93); Robert Oger (1994).

12.3.4.2 *Biometrische Zeitschrift* (*Biometrical Journal*), DR/ROeS

Ottokar Heinisch and Maria-Pia Geppert (1959–66); Maria-Pia Geppert and Erna Weber (1966–68); Erna Weber (1969–88); Heinz Ahrens and Klaus Bellman (1989–95); Jürgen Läuter (1996–99); Peter Bauer (2000–03); Edgar Brunner and Martin Schumacher (2004–08); Tim Friede and Leonard Held (2009–11); Lutz Edler and Mauro Gasparini (2012–14); Dankmar Böhning (BIR) and Marco Alfò (RIta) (2015–19); Arne Bathke (ROeS) and Matthias Schmid (DR) (2020–22).

12.4 Conferences

Table 12.2 summarizes the locations of the International Biometric Conferences (IBCs) and important Symposia by year and location. Also provided are the respective Local Organizing Chair and the Scientific Program Committee Chair, where appropriate. Details of these scientific meetings are in Chapter 9.

International Biometric Conferences (IBC) and Symposia, Details

IBC	Year	Location	LOC[a] Chair	Program Chair
I	1947	Woods Hole MA USA	Chester Bliss	
II	1949	Geneva Switzerland	Arthur Linder	
III	1953	Bellagio Italy	Luigi L Cavalli-Sforza	
IV	1958	Ottawa Canada	John W Hopkins	
V	1963	Cambridge England	Colin Campbell & Michael Healy	
VI	1967	Sydney Australia	HR Webb & Evan J Willians	
VII	1971	Hannover Germany	Berthold Schneider	Peter Armitage (BR)
VIII	1974	Constanza Romania	Tiberiu Postelnicu	Leo CA Corsten (ANed)
IX	1976	Boston MA USA	Yvonne M M Bishop	Edmund A Gehan (ENAR)
X	1979	Guarujá Brazil	Décio Barbin	Geoffrey H Freeman (BR)
XI	1982	Toulouse France	Jacques Badia	Charles W Dunnett (ENAR)
XII	1984	Tokyo Japan	Chikio Hayashi	Daniel Solomon (ENAR)
XIII	1986	Seattle WA USA	Gerald van Belle	Richard Tomassone (RF)
XIV	1988	Namur Belgium	Ernest Feytmans	Klaus Hinkelmann (ENAR)
XV	1990	Budapest Hungray	Bela Gyorffy	Niels Keiding (NR)
XVI	1992	Hamilton New Zealand	Ken Jury	Jean-Jacques Claustriaux (RBe)
XVII	1994	Hamilton Canada	Peter DM Macdonald	Byron TJ Morgan (BR)
XVIII	1996	Amsterdam The Netherlands	Arend Heyting	Susan S Ellenberg (ENAR)
XIX	1998	Cape Town South Africa	Tim Dunne	Geoffrey McLachlan (AR)
XX	2000	Berkeley CA USA	Joan F Hilton	Sylvia Richardson (RF)
XXI	2002	Freiburg Germany	Martin Schumacher	Robert N Curnow (BR)
XXII	2004	Cairns Australia	Kaye E Basford	Louise L Ryan(ENAR)
XXIII	2006	Montreal Canada	James Hanley	Geert Verbeke (RBe)
XXIV	2008	Dublin Ireland	John Hinde	Jean-Louis Foully (RF)
XXV	2010	Florianópolis Brazil	Dalton F Andrade	Vicente Núñez-Antón (REsp)
XXVI	2012	Kobe Japan	Toshiro Tango	Christine McLaren (WNAR)
XXVII	2014	Florence Italy	Adriano Decarli	Brian Cullis (AR)
XXVIII	2016	Victoria Canada	L Cowen & Y Fong	Fred van Eeuwik (ANed)
XXIX	2018	Barcelona Spain	L Gomez & P Puig	Charmaine Dean (WNAR)
XXX	2020	Seoul South Korea[b]	TaeRim Lee	Renato Assunção (RBras)
XXXI	2022	Riga Latavia	Andrejs Ivanovs & Krista Fischer	Kerrie Mengersen (AR)
Symposium	1951	Calcutta India	Arthur Linder	
Symposium	1955	Campinas Brazil	Constantino Gonçalves Fraga Jr	
Symposium	1955	Varenna Italy	Luigi Luca Cavalli-Sforza	
Symposium	1956	Linz Austria	Arthur Linder & Adolf Adam	

[a] Variously called Local Chair, Organizing Chair, Local Arrangements Chair, Local Organizing Chair.
[b] Virtual IBC, because of impact of COVID pandemic.

A

Appendix

Extracts from Cornish and Bennett Letters

Supplementary Materials in Billard (2014) included extracts from letters conveyed to Chester Bliss and George Fisher (Sir Ronald Fisher's son) on the death of Fisher. Details are adapted/reproduced herein.

A.1 Introduction

We append here some details from two letters whose startling discovery in 2011 led to the (Billard, 2014) article. The search for the identity of "Pel" the writer of the first of these letters led the present author, on a somewhat lengthy circuitous and intriguing cross-nations trail, to Oliver Mayo, one of the pall-bearers mentioned in Bennett's letter and the author of the companion paper (Mayo, 2014), and to Henry Bennett. These letters were buried in the IBS archives. To reset the stage, Fisher died Sunday night July 29, 1962, and his funeral was conducted on Thursday, August 2, 1962, in Adelaide Australia. In those days, in 1962, the idea that anyone, be it Fisher's son George or Bliss, could reach Australia from England to attend the funeral was not even a remote possibility. However, as we see in these letters, Fisher's friends, Cornish and Bennett, ensured details were quickly conveyed.

A.2 Alf Cornish to Chester Bliss – 6 August 1962

Cornish – or, "Pel" as he signs himself, an affectionate name used by Fisher himself in earlier writings when referring to Cornish – writes to Bliss with the "tragic news" that "[t]he great man has gone ... one of the great men of 20th century Science." This letter goes into considerable detail about the medical events leading up to Fisher's death on the Sunday night of July 29, 1962, when he suffered a massive embolism in the presence of his surgeon paying his evening visit, dying "within one or two seconds." This was a real shock to Cornish who had seen Fisher only three to four hours earlier when it seemed

Division of Mathematical
Statistics,
C.S.I.R.O.,
University of Adelaide,
Adelaide,
South Australia.

6: 8: 62

Dear Chester,

You have received word of the tragic news of Ron's death,

... ...

I had a harrowing week which reached its climax last Thursday afternoon when I gave a commemorative address at the funeral service. The great man has gone and I doubt whether anyone will develop to take his place - at any rate not in my time. He is certainly one of the great men of 20th century

... ...

Kind regards

Yours

Pal

FIGURE A.1
Extracts from Cornish Letter to Bliss August 6, 1962.

Fisher had "returned to a condition one might expect for someone of his age, 8 days after a major operation."

It had appeared that a few months earlier in 1962, Fisher had suffered some discomforts and so consulted a "medico about the end of May." However, Fisher did not tell anyone of this, not even when Cornish sensed that Fisher was "distrait" in early July. Cornish subsequently learned that Fisher was to have an X-ray in mid-July. "The X-ray revealed a carcinoma at the lower end of the colon." Though "Ron ... made light of it," Fisher "decided to undergo surgery immediately," which surgery was performed on July 21. Carcinoma was confirmed but there were no secondaries revealed; so it was just a matter of waiting for the incision to heal. Infection from his own colon was evident on the 25th; but this was treated, and by the 28th (Saturday) the infection was gone and (Fisher's) "temperature had returned to normal." That afternoon, a small pulmonary embolism developed and was treated. On "Sunday afternoon (29 July), all traces of venous blood accumulation had gone and his respiratory rate was nearly normal and temperature was normal." A few hours later, however, he was gone.

It was a "harrowing week" for Cornish climaxing when he gave a "commemorative address at the funeral service" on Thursday. See Figure A.1.

A.3 Henry Bennett to George Fisher – 5 August 1962

Bennett starts his letter to Fisher's son George with details about "Ronald's funeral" saying that "(i)t was a most moving service and many came to pay tribute to him. ... The six pall-bearers were students with special academic or personal connexions with Ronald ... In gowns and with pride and sorrow, they carried in the casket with his black velvet cap and scarlet robes folded above him. ... ."

Bennett also writes about the medical details and then gives some glimpses of Fisher at the hospital. "He was stoical throughout," "(t)he nurses adored him and he called them his angels," "(h)e was very patient, and, despite sedation, most alert. In his last hours, he was actually working on a Times crossword."

Then Bennett goes on to tell George how happy his father had been "here" (in Adelaide) "with a wide circle of devoted friends," where he "had done a great amount of work," written "over a dozen papers ... important revisions to several of his books," even "complet(ing) a revised edition of 'Statistical Tables' while in hospital." (Fisher) had "been concentrating upon fiducial probability ... He had so much still to give."

The kindly man, described earlier in Section 7 of Billard (2014), is also evident in Bennett's letter. Bennett tells George how his family "adored him and he loved to play with (the children) and we remember him both boisterous and gentle, playing in the garden or telling stories." Bennett promises that "We must honor his memory as he honoured us with his friendship."

Indeed he did; Bennett set up the Fisher Archive in the University of Adelaide Library and issued a five-volume set of *Collected Papers of R. A. Fisher.*

A.[illegible] Henry Bennett to George Fisher, 5 August 1932

Bennett starts his letter to [illegible] [illegible] with details about [illegible] [illegible] [illegible] saying that [illegible] a most [illegible] and [illegible] [illegible] [illegible] to him. [illegible] The [illegible] were [illegible] [illegible] [illegible] [illegible] [illegible] [illegible] [illegible] [illegible] [illegible]

[illegible]

[illegible] [illegible] "there" [illegible] [illegible] friends." [illegible] had done a great amount of work [illegible] paper [illegible] [illegible] Tables [illegible]

[illegible] is also evident in Bennett's letter. Bennett tells George how [illegible] [illegible] [illegible] [illegible]

[illegible] Bennett [illegible] the Fisher [illegible] [illegible] and [illegible] a five-volume set, *Collected Papers of R. A. Fisher*.

Gallery of Secretaries and Treasurers

IBS Secretaries

1947-1955 Chester I Bliss; 1955-1962 Michael J R Healy; 1963-1968 Henri L Le Roy; 1969-1975 Hanspeter Thöni; 1976-1979 James S Willaims

1979-1984 Lyle D Calvin; 1985-1992 Roger Mead; 1993-2000 Elisabeth Barath; 2001-2003 Geert Molenberghs; 2003-2005 Ori Davidov

2006 Ashwini Mathur; 2007-2013 Linda J Young; 2014-2018 James R Carpenter; 2018-2019 Brad J Biggerstaff; 2020-2021 Vicente Núñez-Antón; 2022-2024 Henry Mwambi

IBS Treasurers

1947-1950 John W Hopkins; 1951-1956 Chester I Bliss; 1957-1960 Allyn W Kimball; 1960-1963 Marvin A Kastenbaum; 1963-1970 Henry Tucker

1971-1978 Larry A Nelson; 1979-1986 Jonas H Ellenberg; 1987-1990 Janet T Wittes; 1991-1997 Stephen L George; 1998-2006 Jeffrey T Wood

2007-2013 Linda J Young; 2014-2018 James Carpenter; 2018-2019 Brad J Biggerstaff; 2020-24 Vicente Núñez-Antón

References

Armitage, P. (1996). The Biometric Society – 50 years on. *Biometric Bulletin* 13, 3–4.

Armitage, P. and David, H. A. (eds.) (1996). *Advances in Biometry.* John Wiley, New York.

Bartlett, M. S. (1947). The use of transformations. *Biometrics* 3, 39–52.

Basford, K. E. (2011). IBS: Transforming our governance. *Biometrics* 67, 1185–1188.

Basford, K. E. (2013). IBS: Transformation of our governance. *Biometrics* 69, 300.

Billard, L. (1994). The world of biometry. *Biometrics* 50, 899–916.

Billard, L. (1995). The roads travelled. *Biometrics* 51, 1–11.

Billard, L. (2014). Sir Ronald A. Fisher and The International Biometric Society. *Biometrics* 70, 259–265.

Bliss, C. I. (1958). The first decade of the Biometric Society. *Biometrics* 14, 309–329.

Bliss, C. I. (1965). Visits with the regions of the Biometric Society. *Biometrics* 21, 267–272.

Breslow, N. E. (2003). Are statistical contributions to medicine undervalued? *Biometrics* 59, 1–8.

Cady, F. (1972). Henry Tucker 1923–1971. *The American Statistician* 26, 46.

Campbell, R. C. and Healy, M. J. R. (1963). The Biometric Society; 5th International Biometric Conference. *Biometrics* 19, 668–673.

Cavalli-Sforza, L. L. (1956). The Varenna seminar in biometry. *Biometrics* 12, 93–96.

Cavalli-Sforza, L. L. (1968). Teaching of biometry in secondary schools. *Biometrics* 24, 736–743.

Cochran, W. G. (1947). Some consequences when the assumptions for the analysis of variance are not satisfied. *Biometrics* 3, 22–38.

Cochran, W. G. (1954a). The combination of estimates from different experiments. *Biometrics* 10, 101–129.

Cochran, W. G. (1954b). Some methods for strengthening χ^2 tests. *Biometrics* 10, 417–451.

Cochran, W. G. (1955). Brief of Presidential Address: The 1954 trial of the poliomyelitis vaccine in the United States. *Biometrics* 11, 528–534.

Cox, G. M. (1972). The first twenty-five years (1947–1972). *Biometrics* 28, 285–311.

Dagnelie, P. (1984). 1984 International Biometric Conference Presidential Address. *Biometric Bulletin* 1(3), 3–4.

Dagnelie, P. (1986). Towards greater internationalization of the Biometric Society. *Biometric Bulletin* 3(1), 1–3.

Darmois, G. (1953). Presidential Address: Dignités nouvelles de la Statistique dans la Recherche. *Biometrics* 9, 522–524.

Eisenhart, C. (1947). The assumptions underlying the analysis of variance. *Biometrics* 3, 1–22.

Ellenberg, J. H. (1988). The internationalization of the Biometric Society. *Biometric Bulletin* 5(1), 1–2.

Fertig, J. W. (1984). Biometric Bulletin history (Letter to the Editor). *Biometric Bulletin* 1(3), 2.

Fisher, R. A. (1948). Biometry. *Biometrics* 4, 216–219. [Republished 1964 in *Biometrics* 20, 261–264.]

Gomes, F. P. (1989). The Brazilian Region of the Biometric Society (RBras). *Biometric Bulletin* 6, 25–26.

Lange, N., Ryan, L., Billard, L., Brillinger, D., Conquest, L., and Greenhouse, J. (eds.) (1994). *Case Studies in Biometry.* John Wiley, New York.

Mahalanobis, P. C. (1964). Professor Ronald Aylmer Fisher. *Biometrics* 20, 238–252.

Mayo, O. (2014). Fisher in Adelaide. *Biometrics* 70, 266–269.

Molenberghs, G. (2005). Biometry, biometrics, biostatistics, bioinformatics, . . . , bio-X. *Biometrics* 61, 1–9.

Schneider, B. (1971). Biometrie den 70er jahren. *Biometrics* 27, 264–267.

Index

Note: **Bold** page numbers refer to tables; *italic* page numbers refer to figures and page numbers followed by "n" denote endnotes.

9781032258669